PERFECT PROSPERITY

A Life in Science. A Fortune in Business.

ISKANDAR K. ISKANDAR, PHD

PUBLISHED BY FIDELI PUBLISHING INC.

Copyright © 2015 by Iskandar K. Iskandar, PhD

Perfect Prosperity
Paperback – published 2015 Fideli Publishing Inc.
ISBN: 978-1-60414-877-0

All rights reserved.

No part of this book may be reproduced or transmitted in any form or by any means, electronic or mechanical, including photocopying, recording, or by any information storage and retrieval system without the written permission of the author, except where permitted by law.

"People are often unreasonable and self-centered.
Forgive them anyway.

If you are kind, people may accuse you of ulterior motives.
Be kind anyway.

If you are honest, people may cheat you.
Be honest anyway.

If you find happiness, people may be jealous.
Be happy anyway.

The good you do today may be forgotten tomorrow.
Do good anyway.

Give the world the best you have, and it may never be enough.
Give your best anyway.

For you see, in the end, it is between you and God.
It was never between you and them anyway."

— Mother Teresa

CONTENTS

Foreword ix

INTRODUCTION *An Author's Note* .... *xiii*

CHAPTER 1 An Ancient Tradition: A Modern Dilemma 1

CHAPTER 2 Dignity in Deprivation 5

CHAPTER 3 On the Edges of a World at War 13

CHAPTER 4 The Slums of Cairo … The 'Thief' of Zeitoun 17

CHAPTER 5 Aida and Karam 23

CHAPTER 6 Angels and Demons 31

CHAPTER 7 An Education 39

CHAPTER 8 A Few Simple Twists of Fate 45

CHAPTER 9 Siblings 51

CHAPTER 10 Paradigm Shift. The Nasser Era 61

CHAPTER 11 Home. Grown. 69

CHAPTER 12 Higher Learning: Degree of Difficulty 75

CHAPTER 13 Love, Marriage and Destiny 81

CHAPTER 14	The Six-Day War: An Ordeal of Change.................. 91
CHAPTER 15	The Politics of Religion... 97
CHAPTER 16	Sidestep 101
CHAPTER 17	Coming to America 105
CHAPTER 18	Politics 'Unusual' 111
CHAPTER 19	The Doctor and The Dishwasher 117
CHAPTER 20	Mentors and Tipping Points 123
CHAPTER 21	Of Milestones and Marriages 131
CHAPTER 22	Parting: Children Will Listen 139
CHAPTER 23	Madison: Critical Mass 143
CHAPTER 24	CRREL 149
CHAPTER 25	Earthbound: A Practical Science 155
CHAPTER 26	The Bachelor Doctor 161
CHAPTER 27	The Second Time Around 165
CHAPTER 28	A Matter of Intelligence 169
CHAPTER 29	Publications: A Brain Trust 179
CHAPTER 30	Sugar Blues: The Dance of Diabetic DNA 191
CHAPTER 31	Frozen Ground: The Politics of Energy 195

CHAPTER 32	The Science of Smart Business	205
CHAPTER 33	The 'Tax Man' and the 1031 Exchange	215
CHAPTER 34	A Family Affair: Reunion	221
CHAPTER 35	1990s: Years of Momentum	229
CHAPTER 36	Transition 2000	239
CHAPTER 37	'Sleeper'	247
CHAPTER 38	45 Lyme Road: The Reverse Exchange	257
CHAPTER 39	Evenchance: A Practical Vision	263
CHAPTER 40	Heartfelt	269
CHAPTER 41	Lessons Learned: An Examined Life	281
CHAPTER 42	Next…A New Bucket List	287
MILESTONES	A Chronology of Events	293
	Acknowledgements	303

In loving memory of my mother Aida,
for all she meant to our world.

September 5, 1914–July 24, 1988

FOREWORD

It is a point of contradiction in modern American culture, especially in this Age of Entitlement, that we both honor rags to riches stories and, at the same time, greet them with a kind of skepticism bordering on suspicion. This has particularly been true since the financial collapse of 2008 and the "Occupy Wall Street" movement where people gathered in parks in their Nikes, their iPhones and their Gore-Tex® tents, complaining about exploitation of the underprivileged poor by the rich. (They declared themselves to be the 98%, even though they only had a 30% public approval rating; but counterintuitive revolutions have an arithmetic all their own.)

Most recent case in point of what we like to call the Epidemic Fear of Success: Warren Buffett, "The Oracle of Omaha" [net worth $68 Billion] and one of the three richest men in the world, recently came out with a declaration that it was possible for anyone with $40 in cash to become a millionaire—and that he could show them how to do it.

Of course, the response was an almost universal pushback as something very unrealistic and smacking of Horatio Alger, the author of the very first "rags to riches" novels originally penned in the late 19th Century. Alger's "Luck and Pluck" series of books idealized ambitious young men who rose out of poverty to wealth and prominence. And they invariably accomplished this through diligence, hard work, and their fair share of good fortune (if indeed "good fortune" is fair).

In retrospect, it is accurate to note that the Horatio Alger series of "success" novels fell under assault virtually from the moment they were published. Putting forth a list of personal virtues such as *honesty, thrift, hard work, self-reliance and industry,* Alger emphasized that there were those individuals whose personal code and impeccable conduct lifted them to become part of a "meritocracy." It contained an elite class of individuals whose sole qualifications for acceptance were determination, self-belief and a tenacious work ethic that simply would not be denied. Alger believed that it was this kind of personal initiative and sweat equity that best depicted America as the land of opportunity it had come to represent to the rest of the world.

The fact is that such rare human beings do exist. The truth is that their stories—ones of determination, hard work, and overcoming what seem to be impossible odds—are precisely what make the American dream both credible and attainable. So it is with this in mind that we offer the following memoir of someone who truly rose above his circumstances to become a prominent world scientist and successful millionaire entrepreneur.

His Name is Alex Iskandar. He is an initiator, author, editor and co-editor of myriad scientific papers, more than a dozen books and a series of environmental breakthroughs. What's more, he is an industrious entrepreneur whose vision and drive have helped to shape an entire American community. As such, his is the kind of double-helix life that everyone admires but few could duplicate.

Born into a poor family in Egypt in the late 1930s, Alex Iskandar worked his way out of a caste system that punishes initiative, one that drove him to come to America with a Master of Science degree and $223 in his pocket. He also brought with him an iron sense of purpose that could not be denied—one that led to a PhD in Soil Sciences and a financial and Real Estate portfolio that schools of economics often point to as the perfect business model.

Although he would be the first to deny it, Dr. Alex Iskandar epitomizes the classic success story—a chronicle of personal achievement that both Horatio Alger and Warren Buffet would agree confirms their faith in the sublime destiny of intelligent ambition.

Alex has also done it by applying the simple truth that inventor Thomas Edison once fingerprinted when he said. "People often miss an opportunity because it is dressed in overalls and looks like work."

Dr. Alex Iskandar—a millionaire PhD from the slums of Cairo—recognized that indelible truth a long time ago and has never failed to dress for the job. This is his story.

— *The Editors*

INTRODUCTION

AN AUTHOR'S NOTE

Milestones exist for a reason. For some they are a cause to pause and reflect. For others, they come as a wake-up call that takes place when they reach a certain point in their lives and realize—at last—that they have a legacy they are impelled to share.

That realization is what made my 75th birthday so special, and in a way created a milestone of its own.

Reflecting upon my life, the experiences that brought me some moments of pride, and the realization that I had actually been blessed enough to achieve at least some of my dreams, I decided to host my first-ever family reunion. It was a gathering that included my five children from two marriages, and their children.

Since 75th birthdays are a "Diamond Jubilee" of sorts, I decided to forego any big birthday bash and instead greeted my family with some very simple but specific requests: First, that my children's gifts to me would come in two parts: 1) a letter about what they had learned from me, what they had learned about their own lives and their perceptions of our relationship up to this point; 2) that they set forth a list of short and long term goals that they wanted to accomplish and a game plan of how they would go about it.

I was inspired to do this because of a book and film called *The Ultimate Gift* by author Jim Stovall, and by a series of inspirational stories that have somewhat organically spun out of it.

In this apparent work of fiction, a billionaire industrialist named Red Stevens has just passed away, entrusting his fortune to his relatives who all line up to receive their portion of his multibillion dollar estate.

Red, however, throws some of them a curveball when he picks the most promising among them—his grandson Jason who, Red has determined, has shown some trace of integrity. But the inheritance does not come without a test. To prove his mettle, Jason has to carry out a series of twelve tasks in the form of "gifts" to benefit others and that might, at the same time, come at a personal cost to him. Not surprisingly, the challenges and adventures that Red had set before Jason help to make him a better human being who finishes by actually deserving the fortune he inherits. And though this tale is mythical, the premise made sense to me.

Not that I tried to task my children with a similar set of trials and adventures—but I did do something else. I challenged them to rethink their lives and get back to me with their goals and objectives.

As might be expected in any large family, the responses I received from five adults (ages 30 to 48) ranged in degree and intention from disappointing to amusing to hopeful.

Then it struck me. How fair was it for me to challenge my family with such a list when I had not given them a blueprint of my own life? If I was going to set them this kind of assignment, then I was morally obligated to provide some sort of behavioral model—a perspective, a philosophy and a code of conduct as well. So I decided to do so by writing down this chronicle and getting it into print.

This decision was given added impetus a few months ago when a sudden unexpected bout of heart trouble and emergency surgery infused me with a renewed sense of urgency. It was a scare, but a blessing in disguise because it called me into action.

If I have always believed one thing in my life it is that *Crisis creates Opportunity*. It is my credo. It is something I live by, and it is a course of action I would like to share with you now.

So here, rebooted, is my offer to my children…and to you, the reader: I'll tell you my story—my successes, my mistakes, my learning experiences, my goals and my dreams—if you'll let this work inspire you to do something exceptional with your own.

If you have already achieved all your goals and dreams, wonderful! You can still benefit by using my life journey as something of a validation for your own personal quest. (We never stop learning. I never have. And everyone is my teacher.)

If you haven't done everything you'd like to do, then read on. If there is something in my story that might catch fire in you, by all means read on. If you just want to read an interesting tale that will surely show you some extraordinary places in this ordinary world, then absolutely read on!

After all, we are all on this journey together, and if there is some small "Aha!" moment that I can provide for you—some leverage to success, some way of seeing the world with different eyes and a renewed sense of gratitude—then this book and I will have fulfilled our purpose.

—Alex Iskandar

PERFECT PROSPERITY
A Life in Science. A Fortune in Business.

"Great spirits have always encountered violent opposition from mediocre minds."

— Albert Einstein

CHAPTER 1

AN ANCIENT TRADITION: A MODERN DILEMMA

It seems rare to be born in the midst of poverty with a strong sense of destiny and a true belief that innate human goodness was the driving force of life. And yet it was the driving force with which I always greeted every living day.

I had felt it almost as soon as I could define the world I saw; a sense that there were other realms, a hidden universe that I had yet to discover.

Part of this had to come because of the Egypt that I remember. It was in every sense of the word a paradox. Here we were virtually nestled in the shadow of the Pyramids of Giza and the pharaonic traditions of the oldest civilization in the Western Hemisphere. At the same time, we were brought up in the midst of almost unendurable poverty that would cause many in this modern "high-tech" world to recoil to behold.

Egyptians are what are referred to as Hamitic peoples, which has nothing to do with race and everything to do with language and region of the world. As opposed to Semitic peoples who are not only Jewish but also represent every Arab Nation of the Middle East from Saudi Arabia to Iraq, Hamites are primarily Egyptian.

So, it is a common semantic and sociological error of huge proportions to refer to someone who is against Israel and Israelis as being "anti-Semitic," since you are also talking about Yemenis, Syrians, Saudis, Iraqis and the peoples of Oman and Palestine.

Technically, the people of the Egypt where I spent my early years could be referred to as "anti-Semitic" but in fact that ethnic insult is a peculiarity of the Western Press and was never a term we used in conversation. There was no sense of this kind of division any more than there was any noticeable religious intolerance between Muslims and Coptic Christians (far and away the largest branch of Christianity in Egypt and the throughout Horn of Africa). At the time of my birth in 1938, Coptic Christians represented more than 20% of the population of Egypt, and there was a level of harmony and acceptance enjoyed by both religions—at least to the degree that discrimination by one faith against another was not even a factor, especially to families whose children were schooled together and who saw one another on a daily basis.

The Egypt in which I grew up as a young boy was a de facto monarchy, more specifically a Sultanate that was virtually little more than a British Protectorate. Our young King Farouk was the indirect descendant of the liberator of Egypt, a proud Sultan named Muhammad Ali Pasha. Ali Pasha was a great visionary leader in the early 19th century who emphasized religious and economic freedom for all his subjects and established a strong European style Egyptian state that held its own for centuries. Even today, you can still see the influences of European and Byzantine Architecture in cities like Cairo and Alexandria, and Egypt remains perhaps the most "European" nation in the Middle East (although the Turks may disagree).

In truth, little of that tradition seemed to rub off on Farouk's father King Ahmed Fouad I or on Farouk himself in the 1930s and 1940s, and evidence of the "diluted gene pool" could be noted even then by their lavish and often decadent lifestyles, their many palaces and the fact that little of Egypt's burgeoning economic growth ever filtered down to the people.

In fact, by the time Farouk I had come to throne, Egypt was on the threshold of World War II and was being used as a foothold, along with Libya, for

German Forces in North Africa. Technically allied with Great Britain, France, and the United States, Farouk himself was philosophically sympathetic to the Axis powers (especially due to the king's fondness for all things Italian), a sympathy to which the "Allies" didn't take kindly. Farouk made some pretext at prosecuting Nazis and arrested a large number of German sympathizers and possible spies; but most of it was for show and everyone knew it.

Since I was a mere two years of age at the time the war came into its full expression in North Africa, this had no more than a ripple effect in my life. Egypt was, at best, a pawn in a far greater game of world conflict, and by 1944 our piece had already been played. So that Second World War, raging so fiercely in Europe and the Pacific, had for us become a distant drum. Early in 1945, the British essentially cajoled Farouk al-Awaal into signing an official Declaration of War against the Axis Powers. But by then, the Mussolini Regime in Italy had already collapsed and Germany was in its final throes as a government; so the Declaration itself was merely symbolic.

In truth, the culture of modern Egypt at that time had become a balance of ancient and European influences. The principal language of my native country is Arabic but the currency was, and still is, the Egyptian Pound, obviously reflecting the once strong British economic influence in our country. English is ex-officio the second most prolifically spoken language in Egypt. About 37% of the population has a good command of it. And ties to the West at that time remained strong.

What little I do remember about Farouk's Egypt in the 1940s was that it was a chasm of social inequality between a very rich upper class and a population of poor entrapped in unspeakable slums within eyeshot of the Pyramids of Giza. From the 1930s on Egypt had discovered rich oil reserves that were being mostly mined by well-paid British and American oil field workers, and it produced 40% of the world's cotton harvested and woven by Egyptian tenant farmers and weavers being paid subsistence wages that constituted little more than indentured servitude. By then, the western countries had already convinced Egypt that it was only good for agriculture ["The gift of the Nile"] and could not be industrialized. It was a myth common to that time.

In Farouk I, we had an obscenely sybaritic King (who loved 90 karat diamonds, lived five palaces and ate 600 raw oysters a week). He showed little concern for anything resembling social services but seemed obsessed with making Egypt the show-business capital of the Middle East.

With more than 1000 feature films produced in Egypt in the 1940s, we were the forerunners to what later came to be known in India as "Bollywood."* Egyptian films were quite a popular export in post war Europe and Asia. But it all seemed rather ironic, because very few people in Egypt at that time could even afford to go to the movies. It was a classic case of being led to water but not having a container from which to drink.

---

* In reality, Gamal Abdel Nasser did more than anyone to subsidize and expand Egypt's film industry. But between Western paranoia and the rise of Islamic fundamentalism in the Middle East, many of Egypt's best "movies" vanished without a market, some of them never to be seen again.

CHAPTER 2

DIGNITY IN DEPRIVATION

Our family lived in what can only be described as a state of "dignified deprivation." I know it sounds like a contradiction in terms, but such a thing actually exists.

As Coptic Christians, we had a family tree replete with clergymen—both priests and nuns. As opposed to the Roman Catholic Church that insists upon celibacy as a qualification for the priesthood, Coptic clergymen were allowed to marry, hold property and pass their estates along to their heirs. (The Roman Catholic Church effectively banned marriage for priests in the early part the 4th Century A.D. to eliminate the onus of both nepotism and challenges by heirs to Vatican estates.)

Coptic Christianity is a lesser-known branch of the Catholic faith in that it is neither Roman Catholic nor what is known as Eastern Orthodox. Instead it was yet another spinoff founded by Alexander I, the first Bishop of Alexandria in about 325. The Coptic Church was actually formed as an offshoot of the historic doctrinal split at the Council of Nicaea. Essentially, Coptic Christianity sprang out of the teachings of the Apostle St. Mark and adheres to many of ancient books that have been bounced out of traditional Biblical texts—such as those of Mary Magdalene, Thomas, James the Just, and Judas. Coptics from the Alexandrian Orthodox branch of Christianity were notably persecuted and tortured for their beliefs as far back as the 5th Century A.D. But they held fast to their faith, and a very strong Church with its own "Pope" sprang from the tribulation to endure to this day.

By definition the term, *Coptic,* means "Egyptian." So in its way the schism that took place in the 4th Century A.D. was as much one of nationalism and geographic proximity as it was of doctrinal difference. Having said that, there were many points of faith that set the partitions of Christianity apart…and they were far more than ethnic and geographical.

My maternal grandfather, Father Michael Barsoum, was a Coptic priest, and was reverently addressed as "Father." In Egypt, the Coptic priests are allowed to marry. My grandfather married and sired a large family consisting of six girls and two boys. Young girls at that time were not permitted to attend school or work outside their homes. So the women were brought up to be wives and mothers while the men, at that time, pursued a life in the clergy and other professions. After my grandfather passed away my uncle, Michael Michael Barsoum, replaced him at a very young age as the minister of Anba Barsoum El Erion.

Even though Father Michael Barsoum was highly respected as a clergyman and counselor, he did very little to amass a fortune and more or less left his daughters without the financial support necessary to broker a good marriage. Additionally, as it was considered something of a disgrace for a Coptic priest to send his daughters to school, my beautiful mother, Aida (or Ayda), was denied a formal education. This left what had to be one of the most brilliant and gifted women I have ever known to be functionally illiterate for most of her life. It did not however prevent her from being kind, creative, compassionate and phenomenally wise in ways that constituted a genius all its own. (Gifted, insightful and clever, Aida Barsoum was also inventive. Even though she could not read or write, after her marriage to Karam my mother had a bronze stamp made just so she would be able to use it for signing official papers—a cherished memento that I have to this day.)

My father Karam Iskandar brought good looks and charm to the mix. Born of a wealthy, in fact aristocratic family, he had an appreciation for the finer things of life but had no inheritance to back it up. He was a promising young man, very talented in his way and especially took pride in his abilities as a tailor. Unfortunately, he was not the least bit entrepreneurial and did little to take risks with his profession. Instead, through family connections, he took government jobs to support the family. He was also kind and amiable but flawed in ways that seemed to hurt no one but himself. He was unfortunately cursed with an obsession that I will discuss shortly. And though he struggled with it all his life, it took him decades to finally conquer it.

Still, my parents both exerted a profound influence on my life in ways that perhaps neither expected—and yet from which I learned some of my most valuable lessons as a man.

I, Iskandar Karam Iskandar, was born in Cairo on October 22, 1938. As a first name Iskandar in Arabic translates into Alexander, or Alex. So that was the name I adopted upon coming to America. In Egypt, I was given the nickname Sameer. I was born at home, an apartment on 7 Shouky St., Saem El Dahr, in Cairo. I was the second of ten children, only five of whom would survive to adulthood. Such were the perils of those days in society, compounded by the deadly Rubix of autoimmune diseases that ran through our family.

Most births during this period of time in Egypt occurred at home, unassisted by doctors or nurses and without medical equipment or supervision. Traditionally infants were brought into this world with the aid of a sister, mother, grandmother, aunt or cousin, a female neighbor, or (if the family could afford one) a competent midwife.

I was the second child born two years behind my older sister Eizis (Isis). Although I was too young to remember much about her, my mother often spoke of her rhapsodically as being an extraordinary soul who was both beautiful and brilliant. And when my mother did remember her, she would remark about her namesake, the goddess Eizis, she of Egyptian mythology who became both the Earth Mother and mate to the Sun God Ra. Even in the black and white photos that remained of her, Eizis was exceptionally beautiful, showed an almost angelic aura, and apparently had dazzling emerald eyes, which is a rarity among Hamitic peoples.

The cause of my sister's death was unknown. In Egypt, many children died suddenly during that period from the 1920s to the 1950s. And the doctors simply wrote on the certificate that the death was by "natural causes," which was all they were allowed to note without the benefit of forensic examination.

Autopsies were primitive and expensive in those days; so they were almost never performed. In fact, my sister's death might have been from any number of causes, not the least of which was something that we now call SIDS (Sudden Infant Death Syndrome) that still commonly applies to any child under the age of five.

In truth my family, especially on my mother's side, had inherited a genetic susceptibility to *diabetes mellitus.* And as a DNA roll of the dice would have it, members of both sides of my family were inclined to both *Diabetes Type 1* and *Diabetes Type 2.*

Type 1 is also known as *insulin dependent diabetes.* It is passed down from generation to generation through the DNA. Sufferers of Diabetes 1 tend to be frail and have a limited life expectancy that often does not go past an individual's forties. Diabetes Type 2 is usually contracted in one's forties and is often the result of poor diet, personal stress and or excessive weight gain. It can also strike you down if there is a history of diabetes in your family—unless you are very careful, and most of us are not.

It is a dark tradition that diabetes can be a kind of twisted "gene-pool" lottery, in that some members of the same family are hit with it while others are not. My mother was a diabetic but my father was not. Aida's father, Michael Barsoum, was stricken with a diabetes related neuropathy that was so severe that, in his later years, his right arm became frozen, and he had to perform his priestly duties at mass while carrying the cross in his left arm—something that went directly against the traditions of the Coptic Christian Church. Two of my father's sisters were diabetic, as was my sister and two of my brothers. Eventually, that same genetic strain struck me in my forties. So this vicious and all-too-common autoimmune disease has always has been the Sword of Damocles that has hung over my family.

For those of us in the families Iskandar/Barsoum, the syndrome was both pathological and pervasive. Some of it originally went undiagnosed, but its effects were often dire. The lifestyle for diabetics was especially difficult in those days because the scientific knowledge of its broad-spectrum effects were both incomplete and inaccurate. Until the 1930s insulin was practically unheard of as a treatment; and even then it was considered experimental. When it was finally made broadly available in the 1940s, the injections physicians provided were administered with large pre-boiled "horse" syringes. And practically no one injected themselves. There were no thin, easy-use disposable needles. Trips to the doctors were painful and expensive. And the complications spinning out of this most pervasive of all autoimmune diseases, more often than not, went unaddressed.

Due to hormonal complications and issues of childbearing, diabetes strikes women in far more vicious and pervasive ways than men. And women on both sides of our family were particularly susceptible. There had already been deaths in the family connected to it. So my mother was especially protective of her sons and daughters…to the point of being overprotective.

(As I grew up, I was never allowed to indulge in such normal childhood activities as swimming and riding a bicycle; and to this day I have never learned either. And though my younger brother Mecheal used to sneak out and have his turn at both, I decided to apply my skills and extra time to my studies and improving my mind. There were at times when I resented it, but I realize now that my absence from athletics, in the long run, may have proved a blessing.)

After my sister Eizis died, I became the senior sibling in the Iskandar family. And as I believe there are no accidents in life, it was a responsibility I took to heart at a very early age. I have perhaps always been driven by a sense of fatalism that prompts me to get the most out of every day; I think it hit me early and never left me.

By tradition the majority of marriages in Egyptian are prearranged. Aida and Karam Iskandar broke those barriers of custom. And yet when it came down to actually proprietizing their union, they went through the rigors of parental approval and fortunately received little or no resistance for their efforts. Since both families came from similar backgrounds—and with my father bringing a level of education and social potential to the marriage—prospects for the couple were good in the beginning.

That lasted only a short time because, despite his family connections, Karam Iskandar had become captive to by an insidious addiction to gambling—to any game of chance that happened to be on the table. As a compulsive gambler, my father would take his monthly government salary (always paid in cash) and squander his entire paycheck on everything from dice to a popular Egyptian card game called *RatScrew* (the objective of which is to take all the cards on the table, making it literally winner-take-all).* So deep and broad were Karam's compulsions for games of chance that he would go straight from his payday at his government job and stay out gambling all night, only to come home the next morning without so much as a single coin in his pocket. That would leave him scrambling to get work moonlighting as a tailor so he could feed his family, and it left my mother working at home to help bring in money to put food on the table that father had lost the night before.

Almost invariably, rather than berate my father for his profligate ways, Aida would simply set up her seamstress business inside the house and start generating what little income she could to help keep the family fed until the next "payday." Karam, who was a gifted tailor, would help out in his own way, trying to make amends for his chronic failures at games of chance.

(When I was about twelve or thirteen one of my uncles gave my mother some money, about 25 Egyptian pounds. This was her portion of selling the last piece of land left by her father. My mother used the money to purchase a Singer Sewing Machine, and it turned out to be a boon since both of my parents used that machine for many years to come.)

So, that was what Aida Michael Barsoum had always done; she rose to the occasion. She worked endlessly to get her family out of the very ghetto that hard times and a profligate husband had combined to create. It would be nice to note that this character flaw that so perennially gnawed at Karam Iskandar was something he was able to correct. Eventually my father did correct it, but not until long, long after the damage was done. Despite the trouble all around

---

* Karam Iskandar was also quite fascinated with Backgammon, which had become the rage in Egypt at that time. My brother Saied remembers well that, at the beginning of each month, our mother would send him over to bring our father home from the coffee shop at the beginning of Shoubra Street. And each time she did, Saied always found Karam completely consumed in games of backgammon.

us, my mother always thanked God for everything we did have, and she never complained about our circumstances or cried to others for help.

Times in those early years were tough enough, but with the combination of our family health issues, my father's obsessions with games of chance and the "dogs of war" ever on the fringes of our lives, my first few years were seemingly replete with some terrifying images. Fortunately, I was a little too young to remember much of them at all.

That is the wonderful will of a child—any child—resilience and adaptation. These were qualities I was able to retain most of the days of my life. And many of them were due to my mother who always looked for the silver linings and always made us feel as if we could rise above any circumstance that life might send our way.

CHAPTER 3

ON THE EDGES OF A WORLD AT WAR

Although it is now lost in the fog of remembrance, World War II had a powerful tangential effect on our family and on my life at a formative time…virtually none of which caught my notice at four years of age.

Through some good family connections and political leverage provided by my father's cousin, Ibrahim Abdo, Karam was hired to a relatively high-ranking post in the Egyptian Ministry of Health. So, in 1940, our family moved from Cairo to Port Said—a critical point of entry at the end of the Suez Canal.

Little is remembered about Port Said these days, but for more than a century it was such a hub of commerce, trade routes, travel and intrigue that the writer Rudyard Kipling once noted: "If you truly wish to find someone you have known who also travels, there are two points on the globe you have but to sit and wait, and sooner or later your man will come: the docks of London and Port Said."

Port Said was a highly strategic seaport for Egypt during World War II for several reasons. For one thing, it was on that Northern band that formed the opening to the Suez Canal. So, it was directly in the path of any German counteroffensive toward the objective of seizing Suez. For another, it was a critical point of entry to Egypt from the Mediterranean and was the seat of many important Egyptian government and naval offices, part of the Egyptian Navy and a safe haven for the Allied Fleet. Most important, having Port Said meant controlling the oil traffic that came up through the Saudi Peninsula and the Horn of Africa.

It was right at the height of World War II, but in what amounted to the North African Campaign it was coming to a point of critical mass, and it was taking place just down the road at a place called El Alamein.

It is an often forgotten fact that two of the most decisive battles in the early stages of World War II were fought on the Northeastern Coast of Egypt at El Alamein in 1942. By defeating the Afrika Korps of German Field Marshall Erwin Rommel in the second battle, Britain and the United States established a critical stronghold there that restored British control over the Suez Canal. This in turn blocked the Suez shortcut from the Middle East and to Germany and virtually shutoff their direct fuel chain to Europe. So defeat for the Third Reich from that point on was only a matter of time.

I was too young to remember any of our time in Port Said, but I was told that my father, Karam, was in charge of distributing gas masks to the people of the port city to prevent the inhalation of toxic gases, in the event that this conflict should spread to the East. Among other functions during the war, the Egyptian Health Ministry was also put in place to render aid, safety and, if necessary, sanctuary to people of Egypt should collateral damage from the war spill over onto the civilian population. And for a time it looked very much as if this might take place.

Even though the Allied forces won two decisive victories at El Alamein, there was every indication that the Third Reich might well try yet a third major offensive. So there were many who felt the risk of being the prize port along that critical corridor was not worth the reward of a well-paying Egyptian Government job.

That, among many other things, prompted my father to send his wife and children away from what had to be one of the most dangerous and sought after seaports in World War II and back to the safe ramshackle neighborhoods of Cairo.

My personal recollection of that time in our life was scant. (I was simply too young.) Nevertheless, I do recall a year or two later, finding one of those gas masks my father was handing out, and clearly remember cutting off the black rubber strap that affixed the mask to someone's face and turning it into the perfect band for my homemade slingshot.

It now seems somehow fitting to find relics from a deadly war symbolically becoming mere toys for the next generation—a testimony to life as life goes on and that the worst events in the history are but a passage.

CHAPTER 4

THE SLUMS OF CAIRO ...
THE 'THIEF' OF ZEITOUN

While Karam stayed on to work at his government job in Port Said, Aida took the family to stay with my aunt Moufida in Cairo. By that time I was four and a half, my mother was pregnant with her third child, and we moved into Moufida's small four-room apartment on El Arishi Street, Shoubra where I met my first peer and new best friend, my maternal cousin George Adley.

Moufida, my mother's second youngest sister, was an extraordinarily kind and loving person. And yet, in keeping with Egyptian social custom she was better known to the rest of the world as *Om George,* meaning "the mother of George." Their youngest sister, Liza, was also known as *Om Nabil,* or "mother of Nabil." My own mother was known as *Om Sameer,* or mother of "Sameer"—such was the social clout of male heirs in the traditional Coptic family.

(At the risk of confusing everyone, *Sameer* was my nickname in Egypt before I moved to the United States in 1968. It obviously has more meaning in our language and culture than it does here. So when I came to America, I used something that was more adaptable to this culture. Hence the translation to the first name *Alex.*)

We all lived in this small apartment on the fourth floor at 9 El Arishi. The building was made from limestone and had thick walls, which traditionally took the place of any real insulation. There was no heat or air conditioning in the building. (This was typical of most Egyptian homes at that time.) It was pleasant enough from October through April, but once the summer months started to hit, the temperatures especially in an enclosed neighborhood, would often climb up to 40º Celsius (about 110º Fahrenheit) that, when combined with insufferable humidity, turned every day into a virtual Purgatory.

Moufida's son George was two years younger than I. So when we first arrived he wasn't able to be much of a playmate. A little later in life, however, George and I became fast friends, working, playing and studying together all the way through what amounted to high school.

My mother, my newborn sister Camilia and I lived in the basement of that apartment building until I was nearly five years old. A short time later my father was able to use his family connections to get a personal reassignment from the perils of Port Said back to a lesser government position in Cairo where he rejoined us in Shoubra. But by then the small basement room was way too confining for a family of four. So we soon moved to small house on a farm plot in the El Zeitoun District near what was then considered the outskirts of Cairo.

Later, and for years to come, I would visit my cousin George back at El Arishi during the summer, and we would spend a great deal of time doing what all inventive children with wild imaginations and very little money would do. We would manufacture makeshift toys in the most creative ways and often played games in the street with marbles and a homemade soccer ball. Even then, footballs (soccer balls) were very expensive, especially for children in what amounted to the ghetto streets of Cairo. So the "ball" we actually fabricated was merely an old sock stuffed with rags. George and I also used to save apricot seeds just so we could pretend they were coins or chips that we would bet with, we would play cards and other games to best one another—anything that we could salvage to pull together any diversion for ourselves.

To anyone seeing this from today's perspective, it might seem like a terrible deprivation. But in a way it was a gift, if for no other reason than the fact that it stimulated our youthful imaginations. We had to create our realities, and out of that we learned a kind of ingenuity that children of today are not challenged to put into play. In this age of super toys and electronics virtually out of the crib, today's children are plopped in front of a TV at the age of 18 months and never taught to develop that right hemisphere of their brain.

(In our less technological era, we were not only permitted to do so, our sanity and sense of life demanded it. And I'm certain it helped me later on. To this day, I've never been a "media robot." I've had to use my mind in other ways.)

As was almost always the case with poor children in the Coptic Christian community, the Church became the center of most social activity. The sparse but sparkling rooms in the Community Center and the games they provided were a source of great delight to us; even the simplest facility was like an oasis. So my cousin George and I would spend as much time there as we could, using the Christian Community Club affiliated with my Church of St. Mary (Virgin Mary) in Shoubra, a Christian neighborhood where I spent most of my years in Cairo.

The summer church club had genuine sports with real balls that actually bounced when we used them. Games such as volleyball, Ping-Pong, chess, and backgammon—things most modern children take for granted—were like a fantasyland to us. And as appreciation is often born of a sense of novelty, I have always tried to remember never to take anything or any new experience for granted.

During the summer months, the club was opened from 5 to 9 p.m. six days a week, primarily because evenings were the only time of day that it was tolerable to play as the temperature cooled with the setting sun. When I was growing up, this club had a very positive influence on me, as I was able to interact with my Christian peers and became familiar with Ping-Pong, which I later played in middle school.

During gym class I was able to play both volleyball and basketball but was never avid in my pursuit of them. Games, sports, and other childhood activities were just not a large part of growing up for me. So by the time I reached high school I pretty much quit playing sports altogether. That way I could better focus on my studies (some of this due to the influence of my mother).

As I mentioned earlier, the history of diabetes and other concomitant autoimmune diseases so common in our family caused my mother to be overprotective and prohibited her children from participating in any sports that might constitute a danger to our health. Over the years my mother, Aida, lost four children to illness and what doctors would now refer to as *etiology unknown*. In those days, they called it "natural causes," if losing a child at such an early age is ever a natural thing. And even though my brothers slipped off to clandestine sports activities, I lost my taste for it over the years. And my *sport of choice* became academics.

Originally, our move from Cairo to the outskirts of El Zeitoun seemed like an ideal arrangement since the house we were moving into was owned by my father's two sisters, Oginee and Galila Iskandar. At the time, the change of residence made good economic sense as well as seemingly providing a healthy proximity to family. My aunts were quite fond of Aida, as was my maternal grandmother, Fahima [aka] *Om Karam,* which means (you guessed it) "Mother of Karam."

Nearby, at the end of the short street, my father's cousin Samee Abdo also owned a home surrounded on three sides by a collective of farms. So, for the first time, we not only had day-to-day exposure to close family members, I also enjoyed my first real access to open air and a sense of space. The term *zeitoun* is Arabic for "olive tree." And originally it was a borough far out on the fringes of Cairo where there had originally been a few olive groves.

My early memories of Zeitoun were as a five year-old. Historically, this is a part of Egypt fraught with legend. According to Biblical mythology, it had served as the sanctuary provided for Mary and Joseph when they brought the baby Jesus to escape the infanticidal purges of Herod the Great. And to this day, there is a church located nearby that has become internationally known as the site of the Miracles attributed to "Our Lady of Zeitoun," iridescent images of the Blessed Virgin Mary—a phenomenon beginning in 1968 that influenced so many lives.

Unfortunately, our stay at Zeitoun was a brief one that seemed jinxed from the very beginning—and for a number of reasons.

Perhaps not surprisingly, mainly due to my father's gambling, it didn't take long for the relationship with my aunts to deteriorate. Oginee and Galila got on well with my mother, and they were always kind to the children. Being Karam's sisters, however, they were already quite familiar with his seemingly incurable addiction to games of chance, and they made quite sure to admonish him up front about his responsibilities.

By mutual agreement, Karam was expected to pay at least some of the household expenses from his monthly government salary. He promised he would tow the line, but fell far from doing so. Invariably, every month after receiving his government paycheck (in cash), he would wander in broke from a night of binge gambling, having completely squandered every bit of his money. So, understandably, my two aunts became furious with these self-destructive splurges, especially because he went back on his word. And his inability to bring in any real income soon started to place a financial burden upon everyone.

Then there came the tipping point—The Thief.

Being the natural caretaker she was, my mother often got up in the middle of the night to make tea for Karam's mother, Fahima, to help relieve her coughing (and other complications from her diabetes). Suddenly, as she was preparing the tea, Aida looked up from the stove toward the top of the kitchen door and beheld what at first appeared to be a ghost-like image glaring at her through the glass. Startled for a moment, she tried to scream, but some secret force had completely silenced her. Literally shut off from giving voice to her shock and awe, Aida could only step backwards. And the saturnine man in the window, defiant up to that moment, opted not to break and enter and dashed away instead. Call it divine intervention if you will; call it what you like. But the fact was that the incident frightened Aida so much she lost her voice and couldn't talk for several days afterward.

What is little known about Egyptian culture during that era was that thieves were actually career professionals who had their own "School For Scoundrels," a kind of certification (very much like a degree). Every thief in Egypt belonged to a group or clan in what amounted to a Thieves Guild, complete with a hierarchy and a code of conduct, preset and strict as any set of laws could be.

Aware of this, a friend of the family did some discreet checking up inside the thieves network and received feedback from the perpetrator himself that, by not overreacting, Aida had probably done the singular thing certain to save her life. Had she screamed or become raucously hysterical, the thief assured our friend that he would have been honor-bound to come in and kill her. As it was, she gave him an out. So he took it.

At that point, Aida used that untoward event as an opportunity to save face and at the same time to enable our family to slip out of the toxic environment with these two spinster sisters. She quietly made arrangements to move us all to her family place in El Maasarah, a small village near Helwan—all the while ignoring Oginee's and Galila's pleas that she leave Karam, "for the good of the children."

Still chastened by what they had perceived to be a sacred bit of salvation, Oginee, Galila and Aida herself were all certain that, on that night, God had taken her voice to save her from a certain death. From a modern cynical perspective, such faith might seem far-fetched, perhaps even superstitious. I choose to believe otherwise.

I would come to learn very soon that Miracles do exist—that good and evil are tangible things, and that angels and demons will follow.

CHAPTER 5

AIDA AND KARAM

I had noted that the combination of events—from Aida's dangerous encounter to my father's wastrel games of chance—had precipitated our move form Zeitoun to El Maasarah, a small village near Helwan. Part of it had to do with the fact that Aida needed to put behind her the constant pressure Karam's own sisters were putting upon her to leave him. What I didn't know when all this took place was the fact that she did so despite the fact that both Oginee and Galila had assured her—if she would but take the children and send him away for good—that they would support her for as long as she needed.

It surprised no one that my mother demurred on their proposal and finally refused to discuss it further. Quitting, on anything, was simply not her way of negotiating her world. Above all else my mother was imbued with a generous nature and a forgiving one. Aida Michael Barsoum gave me so many valuable insights into life. One of the most important was that there is a thin line between codependency and Faith; and only truly evolved souls know the difference.

Aida was a woman gifted with a truly elevated consciousness; one might even call it "angelic." I realize that it might come off as a cliché to refer to one's "dear sainted mother." But in this case the seemingly lavish praise comes close to the truth. All I can do is share with you some of the life lessons she taught me, and you may draw your own conclusions.

There were five basic principles that guided her life that she imparted to all her children. And though she didn't formalize them as such, they helped me form some rules by which I tried to direct my own response to every situation: The first was *kindness*—kindness in all things and to all creatures. Aida believed that it took much more energy to be angry and aggressive and that kindness and courtesy were natural aspects of human nature. (And time, at least to me, has proved her right.)

Related to that, she taught me above all else: *Never burn bridges.* "You will meet many people several times along the same path in your life. Let them remember you well. And always keep open the door to reason and accountability."

Hand in glove with Reason, *Faith* was her guiding principle in all things. *"Never lose faith in anything or anyone.* If you believe in people and show them that you do, they at least have a chance to live up to what you think of them. If you judge them too harshly, then there's no place left for them to go, and you will drive them away."

By the same token, Aida believed that you should never give up something, especially if you believed in it. A man would describe this as dogged determination. She would have defined it as *persistence* and would have emphasized the importance of finding one's passion as the ultimate expression of success.

Quietly, almost without notice, my mother was one of the most passionately compassionate people I ever knew. *Her belief in positive outcomes* and her faith in human nature combined to imbue her with a response mechanism that was virtually impeccable even in crisis—especially in crisis. "Everybody faces difficulty. Let that be your test," she would often say. "That's when you get to find out who you really are."

Aida never lost faith, always found the silver lining in every situation, and never failed to give good solid advice on how to make things better. God only knows she had her belief systems tested time and again in so many ways. Sometimes the trials seemed endless and her tribulations, when they came, often hit us in rapid fire.

I have already mentioned the difficulty with money and the stress of my father's addiction to games of chance. Even more deeply imbedded in our family dynamic was the specter of health issues, diabetes and autoimmune diseases that struck down several of my siblings by the time I reached full adulthood. By the time I was five, we had already lost my sister Eizis. And my baby Sister Camilia would not live much longer. So there was an almost palpable sense of foreboding that we were forced to deal with; it was a constant presence in our lives.

Crisis and hardship affect everyone differently, and no two people respond to them in exactly the same way. In our family, my mother and father suffered the loss of their children with a great sense of sadness, and yet they remained strong in spite of it. That alone impressed me as a testament to their love and devotion to one another. Despite all other extrinsic forces that came to press upon them during the course of their marriage, Karam and Aida managed to weather the storm. And there was a core of commitment that overrode everything, one that always held our family together.

They had to have had great chemistry. (They had ten children after all.) Part of the reason had to do with the fact that theirs was originally an affair of the heart and not one made by contract.

One of the traditions of Egyptian culture, especially in those times, was the marriage bond itself. Most marriages, especially among more socially significant families, were prearranged. This was definitely not the case with Karam and Aida.

Karam initially met Aida while she was visiting her sister Moufida's home and was instantly intrigued enough to ask her. He was told that she was unattached and from a respectable Coptic Christian family whose filial ties intersected several generations of clergymen, including her own father and her brother Michael Michael.

My father, was born, Karam Iskandar Takla, on February 18, 1908 to Iskandar Takla and Fahima Gergis Banoub. He was born in Damietta, a Port city about 15 kilometers from the Mediterranean Sea.

Karam was the second youngest child, and the only son of the five children. His sisters (in descending order form date of birth) were Oginee (aka Sania), Galila (aka Mathelda), Hekmat (Om Hanna), and Eilen (Om Mahir). As I noted earlier, Oginee and Galila never married and eventually passed away from complications of diabetes. Karam's sister Eilen also died at a relatively early age. (It was believed she died form some unspecified variation of cancer although no one could say for sure.)

Karam's mother, Fahima Banoub had been born in to a wealthy, socially prominent Egyptian family. Her father was a man of some reputation and stature named Gergis Beh Banoub. The name designation is socially significant because *Beh* is the Egyptian term for "gentleman," just a rank below a *Basha (or Pasha,* meaning "Lord"). So, being named Gergis Beh Banoub would [in places like the United Kingdom] be akin to having a knighthood… or at the very least being referred to as *The Right Honorable* or *Esquire.* At the very least, it indicated someone of noble birth and considerable social stature; so it carried with it a prestige that was unique.

As it was, my father might have eventually inherited a sizable fortune had it not been for a simple twist of fate. Just prior to his death, and for reasons unknown to anyone, Gergis Beh Banoub decided to bequeath all his money to charity, cutting everyone out of his will and leaving his eight children to divvy up the remaining few scraps of his estate that would have, by Egyptian law, gone to the eldest sons.

As daughter # 5 in a family of eight children Fahima would have had little claim on her father's fortune in any case. So now she married Iskandar [Takla] and found that her only tangible legacy was an autoimmune Trojan Horse passed down through the Banoub family DNA. Iskandar Takla, however, did have several family ties to the Egyptian government. So a certain social stature was not entirely denied them.

Nevertheless, by the time the family network of available government commissions had trickled down to his third son Karam, all he had left was some leverage to gain employment as a mid-level bureaucrat and a subsequent lock on job security. I found this to be the ultimate irony with my own father in that, despite his love of gambling where the odds overwhelmingly favored "the house," he was utterly averse to risk when it came to investing in himself, his considerable God-given talents and the chance to be in control of his own destiny.

This contradiction pressed home to me when I was about ten years old. That's when the kindly Jewish haberdasher, Mr. Simon, with whom my father moonlighted as a part-time tailor offered to roll his clothing shop, its prime location and its thriving haberdashery over to him basically for the cost of assuming the materials and the transfer of the lease.

After thinking about it, my father turned him down flat. His reason: he didn't want to risk losing his leverage as a government employee and the security it provided. That's when it struck me that, for all his charm and swagger, Karam Iskandar lacked any basic faith in his own abilities. And that, I realize, is the classic profile of the addicted gambler—that for a few brief hours they can risk everything and blame it on the caprices of lady luck if they fail.

There were rare occasions when my father did win at gambling. Unfortunately, they were so few and far between that I remember every one of them because it was almost like a second Christmas (and took place that infrequently). When it did happen, we celebrated. My father would bring home wine and flowers for Aida, and food and treats for the children; so much so that it seemed somehow surreal. And for a few days the mood in the family would lift to a level that was almost euphoric. This blissful family dynamic was usually short-lived, however, and most of the time we were facing financial instability, a frequent lack of basics, and health that was compromised at several levels.

To my parent's credit they were courageous and supportive of one another throughout so many tragic episodes—especially regarding the ill health and death of several of their offspring. Early on, I remember losing my sister Camilia when she was only seven. It hit them very hard, and yet they seemed to become even more devoted to one another and to my brothers and me (as the only survivors to that time).

So many children in Egypt died in the 1930s, 1940s and 1950s of what the doctors all-too-often resorted to writing off as "unknown causes." Autoimmune diseases such as diabetes, rheumatoid arthritis and epilepsy (all prevalent in our family) were often not accurately identified as such until 1957, and were not even categorized and dealt with as a group until the 1970s. So there is no doubt that some of my siblings—especially my sisters—might have been saved had their conditions been properly diagnosed at an early stage.

As it was, Aida and especially Karam took each of the deaths to heart. And it seemed to me that, with the loss of so many children and with increasing frustrations with mundane government job as an Archivist, Karam resorted to gambling more frequently and with an even more disastrous result. There were so many times that he lost his entire government paycheck that the family and the children would actually not have enough to eat—something that became a dagger to the heart for Aida. Still, she never complained; at least not in ways that we could see.

By tradition, Egyptian families do not always bring their feelings out in the open. Parents do not express all their issues with the children, and they tend to keep their problems as well as their praise to themselves. It is a different show of affection. As children, you know they love you, but they don't always shower affection upon you as they do in places like Italy or Spain or Latin America where all emotions are always out front like window dressing. Still, there is a subtext of strong devotion that is non-verbal and non-physical; yet intense.

It was the same when Karam and Aida would hit a wall with my father's profligate gambling. Most people, especially his sisters and mother, thought Aida was too forgiving of her husband. Modern psychologists would doubtless have labeled their relationship as classically co-dependent. They would have portrayed Aida Michael Barsoum as a textbook *enabler*, and they would have been wrong.

Virtually interchangeable with the spelling of the name *Ayda*, the more modern spelling, Aida, is Arabic for "gift" or "reward." This grew to be a fairly popular name for Egyptian women, especially in the late 19th and early 20th centuries after the initial release of Giuseppe Verdi's namesake Classic Opera, *AIDA* and the critical acclaim surrounding it.

In this tale of Ancient Egypt where greed, avarice and lust for power prevail, the captive princess Aida is willing to give her own life to affirm her love for the Egyptian warrior prince Radames. As such, she comes to epitomize all that is bright, noble, courageous and pure in a woman. In fact, in modern parlance in the Middle East and Europe, calling a woman "Aida," is to describe her as someone who is beautiful *and* smart…as well as imbued with a genuine strength of character. Nowadays, the ultimate put-down—one man to another—is to tell him: "Dude! She's too much Aida for you!" (Meaning: she's out of your league.)

This would have no doubt amused my mother, and she would have accepted none of it. All she knew how to do was to love with a whole heart, to forgive without stipulation and to believe in the healing power of redemption. She never gave up on anything or anyone. And in the long run it paid dividends where my father was concerned. It didn't happen overnight. Some miracles aren't announced with drums or horns; they just come on the soft steps of time and the persistence of goodness. Nonetheless the transformation eventually took place, and our family was the better for it.

Out of his love for his wife (and his family) Karam eventually found God, changed the course of his life, and became something of a master accountant. In doing so, he spent his final years at peace, proud of our accomplishments and dedicated to becoming the kind of giving, caring man he was always meant to be.

Seeing how my mother's faith in her life-mate had been finally borne fruit has made an impact on me at the deepest level of all—that is in believing that you can actually turn anything and anyone toward the light if you can just tap into that secret place that will drive them out of the darkness that's been holding them back. I've put that belief into practice so many times in my life. And though I have been disappointed time and again along the way, I've seen it happen often enough never to give up trying—applying that sacred effort that, as human beings, sets us apart.

I don't mind calling it a faith in God. And I know that will always make some people cringe a bit to hear me say it, but I have seen too much evidence in my life—too many miracles and too much light in the darkness to believe otherwise.

CHAPTER 6

ANGELS AND DEMONS

Our move to Helwan made an impact on my life in several ways, not the least of which was my education; in this case, the interruption of it. Even though Helwan was eventually to become a populous industrial hub in the Valley of the Nile, the village of El Maasarah in the late 1940s was so small that there were no schools near enough to attend. So, during the time we were there, I had to miss at least one grade and nearly missed two. Since "home schooling" as such was unheard of in those days (and since I had a mother who had, herself, been deprived of an education), I had little choice other than to wait.

At such an early age, I might have been tempted to flounder. Instead, being denied access to the books and the opportunity to learn, filled me with what came to be an obsession about being a good student— not only "good," but also the best at whatever course would be set before me. (That was a goal I set for myself, one I was determined to carry through to completion.)

What El Maasarah may have lacked in terms of available education, it supplanted with an exposure to religious and spiritual traditions imbedded in that region for centuries.

One of the earliest Saints of that Coptic tradition came in the 4th Century AD in the noble personage of Mar Girgis. Having lived during the reign of The Roman Emperor Diocletian (known as the scourge of the Christians), Mar Girgis was said to have endured seven years of torture, during which period he died and was resurrected on three different occasions. During the time of his captivity, according to legend, he was visited by Alexandria the wife of

Diocletian who, upon witnessing firsthand the miracles of this mystic, converted to Christianity on the spot. Finally executed one year later on the orders of The Emperor himself, Mar Girgis died a martyr to his faith, and his life and works are still celebrated in the Coptic Church to this day. (Even though the name translates from the Aramaic to "peasant," or "farm-boy," the historical figure, *Mar Girgis,* was somehow transformed in the Roman Catholic tradition into the knightly personage of Saint George—the same Knight of St. George and the Dragon fame…who somehow emerged later to become the "Patron Saint of England.")

On its own, *El Maasarah* was known as the place of *Deir El Anba* Barsoum El Erian, and as such was quite famous. By historical reference, the word, *Barsoum,* means the "son of fasting." Historically, *St. Barsoum El Erian* was born in 1257 to a very wealthy family. The son of a devout Coptic Christian woman, he was not a priest but became the very definition of a "missionary" and what one would have to call a holy man.

Virtually on a mission from God, El Erian was finally sent to prison for preaching Christianity. He lived among the trials and tribulations of what became known as "The Ayyubid Era of Persecution," and ended up becoming one of its transformative figures. In the ultimate irony, the Ayyubid reign was Muslim and its practitioners were direct descendants of Saladin *[Salah El Din]*—Sultan of Egypt and the initiator of Jihad—who, though dedicated to the spread of Islam, was at least pragmatic in his allowance for religious differences. It was perhaps fitting then that a saintly soul such as Anba Barsoum would emerge to overcome all obstacles and bring the comfort of the Holy Spirit to all those who suffered around him.*

According to legend, St. Anba Barsoum was a man of such unwavering faith that God granted him the gift of performing signs and miracles, enabling

---

* One of the articles of faith that separated the Coptic Christians from the traditional Roman Catholic Church was the interpretation of the Holy Spirit. Roman Catholic doctrine teaches that it fills out the Holy Trinity: the Father, the Son and the Holy Spirit. The Coptics believe the Holy Spirit is in everyone, and everything. It is the source of all life—a little too freeform for the Church of Rome to embrace—hence, one of the principal reasons for the original Great Schism.

him to convert even his persecutors and the guards sent to stand sentry over him. Eventually, the governor at that time learned of Anba Barsoum's remarkable deeds and ordered his release.

Afterwards people—Christian and Muslim alike—would come from all over Egypt to receive his blessing. (So much has changed it recent years that it is hard to believe Christians and Muslims once shared many of the same saints, prophets and holy men—including a respect and reverence for Jesus the Christ—but they did. One can always hope they'll get back to that common ground.)

The Monastery in El Maasarah was essentially enshrined when St. Anba Barsoum came there in the early 14th Century to tend the faithful. After weeks of pilgrimage, so many had come from far and near to see Anba Barsoum that the monastery suddenly ran out of food. Mortified, the Abbot confessed his dilemma to St. Barsoum who apparently pulled a "loaves and fishes" kind of miracle out of the kitchen and found a way to feed the faithful for days. And the event seemed to prompt him to settle there for the remainder of his mission in life.

As you might note by now, the name, Barsoum, is a part of our family tradition. So we are the spiritual (if not actual) descendants of St. Anba Barsoum. In fact, my great grandfather, (Father) Michael Barsoum, was a Coptic Priest who became the Abbot there, and my uncle Michael Barsoum became the Abbot after his father died.

I remember even more vividly the challenges and miracles—the struggles between good and evil—that occasionally took place before my eyes. Those came in the form of exorcisms. Because it was a holy center and a place where it was believed "miracles" had occurred, it was also a place where driving entities out of stricken souls was regularly undertaken. So exorcisms were not only recognized but also considered common occurrences, and priests were often sent out from that place as well as other Coptic Christian churches to perform them.

It is a popular scientific platform nowadays to challenge the notion of demonic possession and the rituals undertaken to expunge them. More often than not they are written off as the side effects of superstition in underdeveloped societies. Occasionally, they're looked upon as aberrations of extreme psychosis on the part of the patient or some genetic biochemical overreactions that occasionally challenge explanation.

I'm here to tell you: I have personally come across cases of demonic possession. I have witnessed the exorcisms priests administer to get rid of them. For me, they confirm more than ever that evil does exist. And it is present in a powerful form of consciousness that simply cannot be dismissed as superstition.** When you see someone who is relatively uneducated begin to speak fluent Latin or ancient Aramaic languages and communicate with learned phrases that are both deeply philosophical and eerily perverse, it definitely gets your attention. Even more haunting is the energy shift in any room where the exorcism is taking place. It is utterly noxious and creates temperature variations that are so extreme it is hard to stay in the same room. What's more, the interactions that take place between the exorcist and the possessed can possibly be life threatening for both.

Haunted by the experiences even to this day, I am also reassured by them—that my faith is justified, and that evil exists if for no other reason than to reveal God by allowing us to catch glimpses of his shadow.

One's personal exposure to what seems a lateral dimension is that it creates a reality that is as genuine as it is harsh. And for me, it is this: Just as it is believed that Heaven is hierarchical, so is Hell. There are levels of demons and minions that present themselves in major and minor rankings. I won't try to explain them. But they are there. Egypt—being the oldest civilization recorded before the time of Christ—has compiled a virtual anthology of encounters with angels, demons and Saints. My own father swore until his dying day that

---

** When we were growing up through puberty and even before, several friends of mine and I would actually go to certain locations around El Maasarah where exorcisms were being undertaken by Coptic Priests. We would often sleep on dirt floors in houses of some of the locals and then go the next day to bear witness to a hands-on exorcism. It didn't take long for our intellectual curiosity to be overwhelmed by the physical and metaphysical phenomenon of the exorcism itself. They were both frightening and transformative, and I will never forget them as long as I live.

he had an intersection with such a dark entity that probably changed his life. And it happened during our time around that orb of spiritual energy.

We had only been in El Maasarah a short time when this phenomenal encounter took place.

Having had a falling out with my mother over his gambling addiction, Karam had left the household for a couple of weeks and was coming back to El Maasarah to try and work things out. After getting off the midnight train, he was walking from the station along the country road to our house.

Karam knew the road well enough. It was a dark, solitary trek flanked by cornfields on one side and shadowy buildings on the other. As my father walked along he heard some footsteps behind him. Turning to see who or what it was, he caught the glimpse of a figure shrouded in a long coat following him. Even though this mysterious fellow made no overt threat, he did quickly catch up to Karam and started walking beside him stride-for stride.

Looking down in the half-moon light, my father started to realize that this "man" had no shoes on his feet and was walking on cloven hoofs. (Not quite believing his eyes, he looked time and again to make sure.)

One warning every child in Egypt is taught to heed is that, should you confront demons or dark minions in the flesh, you should never look them fully in the face; because if you do they will seize your vital force and make it their own.

Brushing up against this hideous apparition walking next to him, Karam at once remembered admonitions from his own father challenging him about his life's path... "Have you made any *hageb* (contracts) with the Devil? If so, they will come back to haunt you."

Anticipating the worst possible outcome, my father kept his head down and would not engage this entity in any way for fear of being cursed or worse—possessed. To counteract this, Karam prayed. He prayed every prayer he could imagine—to Christ, to the Holy Spirit, to the blessed Virgin, to God Almighty and to all the angels on high. He promised if he were to be delivered from this diabolic encounter that he would amend his ways. And it must have worked because in a few minutes the "devil" slipped away as if he had disintegrated.

My father wasn't sure about the exact moment. All he could remember was this: it had seemed a warning to him that he needed to rethink his life, and to understand that darkness—true Darkness—is an absence of God. That is when creatures come in the night, and that's where he had been.

My mother always felt a closer contact with the angelic side of things, and grew even more prescient as she grew older. (My nephews, my sister Fayza's sons Maged and Nader related many occasions later in her life when Aida was actually able to predict events before they took place. I believe it, since she always had a way of seeing positive outcomes in the worst of times, but never gloated when all the good things actually came to pass. They were part of the course of things in the way she beheld her world.)

As it was, my mother was especially drawn to miracles and miraculous occurrences as proof of God's love, not the least of which took place shortly before I left Egypt—in 1968. I refer to the shrine of what has come to be known as "Our Lady of Zeitoun." And it had to do with a series of mysterious appearances by The Blessed Virgin Mary just above the cathedral there.

The first sighting took place on or about April 2, 1968 when a Muslim bus mechanic named Farouk Mohammed Atwa, who worked across from the church of Saint Mary, thought he saw a woman attempting to leap from atop the steeple.

Two of Mohammad Atwa's coworkers also clearly caught sight of the "ghost-like" figure at the same time, and so they all decided to call the police.

According to the original police report, this paranormal sighting was merely a reflection of the light from the street lamps in that vicinity. But by then a crowd had gathered. Word spread that this Vision of St. Mary the Mother of Jesus had appeared, and droves of the faithful would not be dissuaded.

About a week later, a very similar manifestation took place and continued to do so sometimes two-three times a week, for several years.

During the three plus years this phenomenon was taking place, the Coptic Christian Pope of Alexandria, Kryllos VI, appointed a committee of high-ranking clerics to investigate all aspects of it. And in relatively short order, this college of priests and bishops came up with an entirely different interpretation: that this series of appearances was in fact a divine manifestation and therefore a miracle. Needless to say, word spread around the Christian world all the way to the Vatican where Pope Paul VI was compelled to send his own ligation down to the site, all of whom came away with the same opinion.

Finally, the vision of "Our Lady of Zeitoun" ceased altogether around the end of 1971—three plus years after the original event—but not before thousands of people had borne witness to it. One of them was my younger brother Saied who remembered that the vision he witnessed actually stayed in place for several days in a row (seemingly to confound the skeptics and validate the sightings of the faithful).

One can question the appearance and look for scientific explanations all day long. But faith will have its say. From that time and for the next three years, pilgrims came and brought their ill and infirm, their deformed and dying to the shrine. Miracles were performed. People were healed. Diseases were cured. And madness was abated. Who is to say why, other than the fact that each of us holds that miraculous source within us? But the Shrine of Our Lady of Zeitoun portrayed a spiritual clarity all its own…one that has survived to this very day.

For the next several years after the "psychokinetic occurrence," police and scientific investigations of the site at Zeitoun netted absolutely no explanation for the phenomenon. No device was found within a radius of fifteen miles capable of projecting the image; nor did anyone come forth with confessions of "manipulating the event."

My mother came to the hallowed ground on several occasions and believed with all her heart that the miracle she beheld was real. And she was not alone. So did (then) Egyptian President Gamal Abdel Nasser, numerous TV networks and members of the Egyptian press. Finally, having been unable to produce an alternative explanation for the luminous sightings, the Egyptian government accepted the apparitions as valid.

I realize, even as I relate these events that about half the people reading this will approach the subject with a sense of disbelief. I have to admit I have always found that a bit odd: that we as sentient beings will accept and even embrace the dark sides of human nature—wars, disasters, starvation, abuse, crimes against humanity, and lack—simply as the way things are. And yet these same people will look upon a gentle angelic nudge such as a miracle, a vision or a healing as incredible. To me, such things are as natural as breathing, and part of what makes this journey called life worth every single step.

CHAPTER 7

AN EDUCATION

Cut off from many of the basic amenities and without viable schooling for the children, Karam and Aida moved us in 1946 from El Maasarah back to Cairo. Somehow by then our family dynamic had changed, at least temporarily, for the better. Chastened by his experience with the dark entity and the prospect of nearly losing his wife, Karam Iskandar had grown a bit more stable and actually started to settle down. (At least it seemed that way on the surface.)

Immediately upon our return to the City, I was allowed to enter the mainstream Egyptian educational system and was immediately enrolled in a *Dar El Tarbia El Hadietha* School in Soubra. The *Dar El Tarbia* classification loosely translates into "Modern Education," in English; and it was the official designation for schools within the province of the Ministry of Education. Whatever one could say about Egyptian social services at that time, the educational system was in every way a meritocracy. By definition, it mandated that children would be able to advance in direct proportion to their own abilities and efforts all the way through college.

What this meant to children at the "grade school" and middle-school levels was simple and, at the same time, pretty intimidating: Children who made good grades and met specific criteria for achievement advanced to better schools and better career opportunities later in life. Those who did not would be relegated to lesser schools and experience limited chances for advancement. Allowing for late bloomers and a conversion in consciousness may seem logical and a matter of common sense here in America, but in Egypt there was little flexibility. Unless you were laser focused on your studies out of the gate, your career in academics was virtually doomed. So the process of elimination for underachievers or "average students" was practically Darwinian.

Due to our El Maasarah hiatus, I came into the First Grade at the age of seven. That made me nearly two years older than most of the children at that level, and I still remember feeling both vexed and motivated to prove myself from the moment I set foot in class.

By holding back for two of the most formative years of my life, I was like an arrow cocked in a bowstring just before it is fired—tautly drawn and pointed straight to the target. Especially since I was a year and a half older than the other children, I had developed the need to prove a point—that I had both the focus and maturity to be the best. I applied myself with extra effort in everything; and my grades always reflected that commitment.

I credit my mother for much of this. Denied an education, Aida was passionately committed to seeing to it that her children enjoyed every advantage when it came to higher learning. By the same token, she was often so protective of us all that she was entirely risk-averse where our advancement was concerned.

I can still remember the school and its principal as if it were yesterday. The school was located on Badea Street at the corner of Maasarah Street. It was a small rudimentary building, but to my seven year-old sensibilities it looked like a coliseum. Consistent with Egyptian tradition at that time, children attending classes observed a strict dress code, and every morning we would all line up for attendance.

As part of my school "allowance," my mother used to give me a coin each day, which was equivalent to ½ of a US cent—one-half a "Lincoln-head" penny—and that was it. (It seems virtually inconceivable now, but these coins actually had purchasing power. So, I would save up for a few days so that I could buy a coconut, a mango, an orange, or a piece of sugar cane toward the end of the week. I would do this so I could take it home as a surprise to share with my mother and my siblings.)

The principal at Dar El Tarbia was a lovely woman named Miss Emily who dressed well and always looked a bit aristocratic for her station. She was also a Coptic Christian who had a keen insight into the potential of the children who attended her classes. That instinct came into play for me by the time I was ready for my second year. Having evaluated my grades as well as my interaction with the other children along with the fact that I was more distinctly mature than they, Miss Emily determined that I could easily skip that year and go straight into the third grade. To her way of thinking, that would have enabled me to make up the year I had lost and bring me to a level more suited to my abilities. But to Miss Emily's surprise (and mine) Aida insisted that I remain at the current grade level. Despite the pressure to promote me, she would not budge on the issue.

My mother had "an instinct" about such things, and eventually she may have proved to be right. Because I was older than the children in my class, I didn't share in their games and activities. They didn't interest me, and so I sought the sanctuary of my studies. That in turn made me an even better student; it was my social "anchor" at the time.

During the first couple of years back in Cairo our family rented an apartment near the El Hadietha School. [This was located at 40 Harrett Gad Sheriff, Cairo. Although now a mere shell of a building, it still stands to this day.] It consisted of a single bedroom on the first floor with a small play area and a bathroom. And though it seemed secure for a time, any illusions we might have had about my father's "reformation" were very shortly dashed.

One night, after we had been living in these small apartments for a few months, some strange men came in and hauled away all the furniture, including three prized oriental carpets. Later I came to learn that my father had lost all his money gambling and had borrowed more by leveraging the only assets he had left. Since it soon became evident that Karam was going to default on his debt, these "strangers" were actually enforcers his creditors had sent along to exact payment—in this case, our furniture and carpets.

During my early to mid school years, we lived in poverty. My mother, my father, my siblings and I moved into in a single room within a three-bedroom apartment, shared with three unrelated families. This was located on the same street *(46 Harrett Gad Sherrif),* two houses up from the other one but with even lower rent. Each family had two or three children. We had no electricity but we did have unheated running water. The "sewer" was actually just a holding tank—not even a septic tank. Because there was no infrastructure we had to be very careful about the amount of water we used and exactly how we used it.

My mother cooked all of our meals on a small kerosene stove, like a camp stove, in our room. We had no refrigerator. We had nowhere to keep leftovers or store perishables; so shopping for food had to be a daily experience. There was, however, an old fashioned stone oven under the stairway that Aida used to bake bread for all three tenants.*

Our diet consisted primarily of vegetables and rice, and yet my mother was able to prepare them in ways that always made them seem varied delicious. We couldn't afford to buy meat or fish except perhaps once a month or for special occasions such as Christmas and Easter.

(Christmas in the Coptic Christian faith is observed with a level of emphasis that is longer term and a bit more intense. It doesn't fall on the same day as traditional Roman Catholic version and is observed with 30 days of fasting followed by 30 days of gifting, feasting and celebration. [Given our degree of lack during the time I was growing up, my family had the "fasting" part down pretty well. The gifting and feasts were a little harder to come by, and yet I remember them with great fondness as times when we pulled something extra out of ourselves.] We found ways to love and share and be unselfish. And that generates a sense of gratitude that money cannot buy. On those Christmas occasions, Aida would manage some special gifts—usually things the children really needed such as a shirt or a new pair of shoes. We never had the luxury of

---

* It was here at 46 *Harret Gad Sherrif* that my brother Saied was born and spent the first few years of his life sharing a unique experience: One of the tenants, a lovely Muslim woman named Om Lilia had the largest bedroom with a balcony all to herself. Saied used to wake up in the middle of the night, go to Om Lilia's room where she used to breast-feed him and put him to sleep in her bed.

toys or games. And because this was a Muslim Country, Coptic Christians even then never flaunted their "holidays.")

After three years of grade school, I was transferred over to the adjacent school. The school had the same name, *Dar El Tarbia/El Hadietha,* but had also been extended to include middle school grade levels. As I was starting to develop physically, I ended up in middle school playing on the Ping-Pong team; and in my last year I became Team Captain. Although we never quite reached the Chinese or Korean level of world class "contact Table Tennis," we were still pretty good. With four players, we practiced and played other schools in that district of Cairo and managed a respectable record.

Ironically, despite their support in so many other ways, I can't recall a single time when either of my parents attended one of our matches—an indication of the low-value that our family placed upon sports.

Academics were entirely another matter. In the National Exams at the end of middle school and high school, I made outstanding grades and was ranked in the top 2% of all the students nationally tested.

I have customarily been an early riser, and always used that time to pursue my studies. Originally it was probably due to the fact that bed bugs seem to be more active just before dawn; and, in those days, they were my alarm clock. Their bites would rouse me as a kind of rebuke for my staying in bed. I would get up and study while everyone else was still asleep. It was much quieter at that time of day—still dark for the most part. So I would either study by a very small kerosene lamp or would go onto the roof for the natural light and fresh air. And I remain convinced that the human mind is at its most facile and receptive when it is fresh from a good night's sleep.

My parents never had to ask me to study. I used to feel guilty if I did not get excellent grades. Guilt can be a powerful motivator if used upon oneself to achieve something extraordinary; and it must have worked when I used it on myself because I was always at the top of my class.

I knew that always meant the world to my mother. But what surprised me most was also how much it meant to my father. On several occasions over the years I would wake during the night to find Karam Iskandar pouring over my books to see if I had done my homework…and later reviewing my grades on quizzes and exams to see if I was keeping up the high standards I had already set for myself.

This gave me a greater insight into the man in that he did this quietly, never trying to pressure me or assert his persona in patriarchal ways. And yet I could occasionally detect a glimmer of pride in his eyes—the kind a man reveals when he has found a place to put his lost dreams, and a son with an intention to carry them through.

CHAPTER 8

A FEW SIMPLE TWISTS OF FATE

During my transitional years from elementary to middle school, our family lost three children—all younger than I—and each loss struck us in different but deeply felt ways.

My younger sister by two years was Camilia. She was a sweet lovely girl—not as brilliant perhaps as Eizis had been, nor as beautiful, but with a gentle spirit all her own. She was never in the best of health; so there were always concerns. Still my mother was shocked when Camilia was suddenly stricken with Diphtheria, and she felt guilty about not having been able to catch it in time. (And yet who could hold her accountable? The medical diagnostics of those times were so limited, especially for the poor, it was a small miracle that anyone survived to adulthood.) As it was Camilia died at the age of six, and of a disease that is easy to cure today.

Later, during my middle school years, my younger brother George died of cancer around age 10. And within 40 days of George's death, I lost a baby sister named Nadia who was a mere two weeks old. Sudden Infant Death Syndrome (SIDS) was not even an acknowledged pathology until the 1970s. And though "crib deaths" were common, the lack of information about causality placed such a burden upon the parents—especially upon the mother who was usually still nursing—that it left entire generations of women emotionally broken to have to bear the guilt of such a loss.

That guilt, for the whole family, was doubtless exacerbated by the death of my brother George, who was hit with a terminal prognosis virtually over night. Up to the age of ten, George had seemed so very robust that the onslaught of this (unspecified) cancer struck us like a bolt out of the blue. And, as such, it seemed to be both vicious and unjust. The matter was made even worse when the final determination was made and the attending physician—my father's own cousin Ramzy—recommended giving my brother George a fatal injection that would end his life and put him out of his misery.

Karam Iskandar was not a violent man, but the prospect of his own cousin offering to put his son down as if he were the family dog was almost too much for him to bear. I thought my father was going to punch the man out. (And I wouldn't have blamed him in the least.) Instead, he ordered Dr. Ramzy out of the house and refused to speak to him again for almost a decade.

Nadia's death 40 days later felt like the Book of Job. I became convinced that we were the cursed of God. And yet, like Job himself, our faith was the only thing that would lift us from this.

Aida Michael Barsoum was a strong woman, but the impact of these two deaths of her own flesh and blood inside of two months time was too much even for her to bear. Understandably for several weeks, she became inconsolably depressed. Then, with renewed determination, my mother completely rallied; somehow the adversity had forged her in stronger steel.

When I was older, Aida later confided that, during that time, she almost lost her mind. The death of three children—George, Camilia and her beloved Eizis at such an early age—caused her to feel especially cursed and in dread of life itself. (Then the crib death of baby Nadia sent her completely over the edge.) Only her faith in God and the support of her husband helped her weather the storm and even strengthened their devotion.

Tragedies, as we get older, carry more gravitas. They never seem to strike a child as hard; children are simply more resilient. And yet this string of terrible events gnawed at me with a metaphysical dilemma that I have never entirely been able to resolve. If there is kind and loving God who "has His hand on us all," how does one resolve the departure of such sweet and beautiful souls as my brothers and sisters at such an early age? Is there a Divine Plan? Or are we just set down in this time and space to fend for ourselves, subject to some vicious Cosmic Lottery where winners and losers are merely grist for the Mill?

If anyone is living testimony for the force of a Higher Hand in their lives, I am that witness. There were many times when I should have died but didn't, when I could have come into harm's way but escaped it, when I could have fallen victim to tragic circumstance but came out with a phenomenal stroke of luck in its place.

If cats have nine lives and we have Guardian Angels, I certainly needed my share of both when I was a child. My flirtations with "terminal velocity" started out when I was barely six when my cousin Malak Labib (aka Bahlool) tossed me over a second-story railing. Malak was about eight at the time, and so enjoyed a sizable advantage in size and weight. So when he shoved me with all his might, I tumbled like a shot straight down on my head into a large cooking pan. The blow, like the tolling of a bell, sent everyone rushing to look and put a gash in my head that left a scar to this very day. (Those who witnessed the event at the time were convinced that the fall might well have killed me. And had I struck the pan at another angle, it might well have broken my neck.)

My second near death experience came with a bout of Diphtheria—the very syndrome that killed my sister Camilia. Diphtheria is an upper respiratory tract disease that is both deadly and contagious. It can strike anyone but is most dangerous to children under 7 and adults over 70. It reveals itself through esophageal ulcers and swollen lymph glands, but by the time the condition advances to that level of visibility it is very often too late. Diphtheria carries a pernicious bacterium that moves through the human system with sonic speed, and once it hits a certain level, nothing will work to stop it.

That was the deadly peril I faced right after I had turned seven. At the time, I was enjoying some sweet red dates, a treat that Karam had brought home for all of us to share. After eating a couple of them, I started to have problems swallowing. When the swallowing grew more painful, I complained to my mother. She opened my mouth with a spoon to look at my throat and it struck her—she had borne witness to this very disease and recognized its symptoms as both deadly and in urgent need of treatment.

When she told my father, he immediately rushed me over to see his cousin, Dr. Boutros Yousef. Boutros was in medical school at that time but had certainly had enough training to diagnose the condition—and recognized immediately that this was indeed Diphtheria. Without a moment's hesitation, he wrote a prescription for my father to get to the pharmacy and purchase an antibiotic shot. Dr. Botrous injected me right away...and later told my father that if we had waited, even a few hours longer, I too quite probably would have died.

By acting as quickly as they had my parents had saved my life. And thank God that Botrous Yousef was on hand at precisely the moment he was needed.

On another occasion, when I was about ten years old, my own actions almost brought me down and yet paradoxically may have saved me as well.

Having always taken great pride in my academics, I was particularly honored when my father had saved up enough money to present me with a refillable Schaeffer ink pen as a special holiday gift. It is hard to believe now that, as recently as 1950 most students in countries like Egypt were still using pens with inkwells on the desk where you had to dip the tip in the ink and then wipe off the excess on a blotter even to get it to write. Refillable pens were like rocket ships—they were yesterday's iPhones—and this Schaeffer siphon pen was the very state of the art. Even to have one marked someone as both learned and well-to-do. And for a kid my age, this had to be the ultimate status symbol.

I knew how much time and effort it took for my parents to come up with a present like that. It was beautiful, marbled and gleaming. So my pride of ownership overwhelmed all logic, and I displayed it in my front jacket pocket, which—in the Egypt of those days—was like sticking a "Steal me" sign on it.

A couple of days after flaunting this instrument, I was walking down through a narrow alley near our apartment and got brushed by an older man who came way too close for comfort. After a moment or two, I looked down at my jacket pocket and immediately realized that my pen was missing. Turning back to look, I could see that the man had not gotten that far along the passageway. So, without thinking, I dashed over to catch him.

Egyptian men at that time—especially farmers, vendors and thieves—often wore large dark blousy shirts they called a *gallabia*. Gallabias are noted for their bulky flowing sleeves that are large enough to hold just about any small object without detection. In modern times, they are occasionally worn by women as a kind of fashion statement. But this man's gallabia was definitely a thief's tool to pick the pockets of unwary fools like me.

Whatever I might have lacked in common sense, however, I more than made up for in what the Jews refer to as *Chutzpah*. I immediately ran up, yanked on this man's sleeve and demanded that he, "Give me back my pen!"

By now, you are familiar with the tradition of Egyptian thieves, their code and their culture. It would not have been out of the question for this man to have killed me. Instead, something about my 10-year-old's bravado amused him. So he pulled my pen from his sleeve and gave it back to me.

Completely wrung, my courage exhausted, I ran home and told my mother who both scolded and embraced me for such a bold and foolish act. When she told Karam, my father assured me that I was born under a lucky star.

"He might have beaten or even killed you. Nonetheless here you are." There I was indeed—both lucky and blessed.

CHAPTER 9

SIBLINGS

L ife can be relentless. It is a celebration of itself, and if we are to participate in it at all, we should recognize its cruel brevity and at the same time our opportunity, even our responsibility, to make the most of it. We should also never forget those that this shared journey of life has somehow removed from our presence—because in so many ways they become both symbol and sacrifice that the torch has been passed and we should carry on.

Families are of course our closest points of reference. Parents are our role models—for better or worse—and the marks they leave upon our character are indelible. I have also learned that each family member—every brother and sister—brings his or her own threads to the fabric of the family dynamic. Time is not necessarily the only factor in considering this. Some carry a spirit into this world that endures long after they have left it. I have ten such examples in the family Iskandar.

Initially, I mentioned the child first born to our family, my sister Eizis. Even though I barely remember her, the impact this beautiful soul made upon my mother remains etched in my memory forever. For as long as she lived, Aida celebrated her almost angelic qualities. And for a time I admit that I thought Aida favored my departed sister. Later I came to understand that each of my mother's children were very special to her, and she treated the loss of every child as if a part of her heart had been torn away.

Had she lived, Eizis would have been the senior sibling and as such would have acted as a second parent to all the children who followed. Because she crossed over at an early age, that mantle of responsibility fell to me. Perhaps because of the inept medical practices of the time, I lost both my sisters (Eizis and Camilia) when they were very young.

I'm sure that Camilia would have been saved had she been stricken with Diphtheria a few years later. I was lucky. Some of my other siblings were not. My younger brothers Mecheal and Atif both turned out to be diabetic. Somewhat due to my discovery of my own Type 2 diabetes around my 40th birthday, they were able to get tested and have the pathology revealed.

My brother Mecheal, born seventeen years after me, was both active and rebellious. I mentioned that he used to defy my mother's prohibitions about swimming and cycling and would slip away to participate in every sport he could.

My younger brother Atif was a bit less physical but was very bright and had inherited our father's people skills. He could always work his way into, or talk his way out of any situation he liked—a facility he still enjoys to this day. (My mother used to say, "When Atif was born they pulled him out with his tongue.")

All of the Iskandar children were good in school. Everyone achieved high marks. (Perhaps not as high as mine, but quite good enough.) I credit much of this to our mother Aida. She was intent on seeing to it that we seized every educational opportunity presented to us; and for the most part we did. Mine is a family replete with Bachelor's and Master's Degrees—and some PhDs, as well. My brother Saied [10 years my junior] managed over his career to accrue no less than four degrees—a Bachelor of Science, Master of Arts and two MBAs.*

Noting my mother's dedication to our scholastic accomplishments, I should also credit Karam Iskandar for his high intentions as well. Having

---

* Of course not all achievement is measured by "degrees." I'm happy to say that all my brothers have gone on to become successful in their relative fields, and I'm very proud of them all. Having said that, I have also come to realize that life is not always fair. Education sooner or later becomes a means of social leverage. It gets you jobs. And in certain fields it even gets you published.

been well educated himself (with an "accounting" degree), he saw education as the pathway out of penury and into a fully rewarding life.

He certainly encouraged me at just the right times to pursue my studies and secure a degree. And later, once he stabilized his own life and won his bout with his "demons," he seemed to double-down on making sure that my younger brothers and my sister made something special of their lives.

My younger brother Saied especially noted that, by the time he started coming of age, Karam had taught him how to prioritize most problem-solving issues in his life—especially when it came to business challenges.

"Always work backward to find the solution," my father always insisted. Applying an old accounting principal, Karam taught Saied to see the perfect result first. "And then reverse the steps that come from result to initiation. That way you're sure to leave nothing out." Good advice, and Saied said it was something he followed with great success for most of his career.

In fact, it was an extension of that belief system, and the expertise that came out of it, which brought Karam Iskandar to be one of 5 key individuals chosen to reconfigure the Egyptian Government's official "Accounting Department" in Tahrer Square in Cairo. Later, sometime in the early 1970s, this department changed its name to "Central Government Accounting for Egypt" and moved to Nasser City where it now employs thousands of people.

It was also in the early 1970s that our father co-created a non-profit charity called the "St. Paul Orthodox Coptic Meeting." Encouraged to do so by Aida and by his sister Galila, Karam followed suit by committing himself wholeheartedly to "the Cause," consistently helping them balance their finances while also handling any audits that so frequently accompany the Egyptian Government's scrutiny of nonprofits.

Needless to say, it is a rare family that brings a perfect set of circumstances into this world. In our case, the family Iskandar had a double hit of financial challenges and health issues.

The disease matrix that chronically carried itself into our DNA on both sides of the Iskandar/Barsoum gene pool made a much more savage mark on the women in our family. Unfortunately, this is not at all unusual. Autoimmune diseases such as diabetes and rheumatoid arthritis strike men and women in approximate equal numbers but apparently create more side-effects and complications in women than men; ones that are far more pervasive in both effect and duration. This was certainly true of my sister Fayza who turned out to be the only daughter in our family who actually survived to adulthood.

Very much endowed with her mother's gentle nature, generosity of spirit and maternal instincts, Fayza was a lovely girl who was cursed at an early age with a crisscross of not one but two autoimmune diseases—diabetes and crippling rheumatoid arthritis. The diabetes was something the doctors caught early. But the rheumatoid arthritis didn't strike her fully until she was about twenty—right in her prime as a young woman. From that point on it limited her ability to perform certain activities, and her condition could only be ameliorated by medications that she was compelled to take for as long as she lived.

Fortunately, this horrendous genetic imbalance didn't dampen Fayza's spirits or keep her from being a great sister, a warm affectionate family member and a phenomenal friend. Since Fayza was just two years older than Saied, those two in particular grew up side-by-side, remaining constant companions who shared a trust and mutual respect that carried throughout their lives. This strong filial bond between brother and sister also filtered over into their respective families (to such a degree that Saied brought his new bride over to Fayza's to spend a few days of his honeymoon with his "favorite sister").

Later when Fayza was in hospital after having given birth to each of her two sons, our mother had already become too frail to visit her as frequently as she might have liked. So Saied often acted as messenger for Aida and the bearer of gifts and cards from mother to daughter.

The indelible imprint of Fayza's warmth and generosity of spirit has always stayed with me with memories that become more poignant as the years wear on. I shall always recall how often she looked after her brothers even when she was a teenager, doing so in many ways that virtually went without notice. I still remember one such instance that forever touched me deeply—so pure was her generosity, so very strong her heart: Grapes were always a special treat for us when we were growing up. And whenever we got clusters of them from the local market, Fayza always partook of the smallest grapes first (the youngest and sourest of the lot), making it a point to save the largest and sweetest for her brothers. Knowing how acrid they could be, I one day asked her, "why?" And Fayza, as only she could do, sweetly acquitted herself with a lovely little white lie.

"Oh, I prefer them," she announced to me. "The big ones are just too sweet."

I wasn't buying for a minute, mind you, but then I realized that rather than call her out on this. I would simply let the issue drop.

Being around Fayza, even for a little while, made you quickly aware that she had inherited Aida's loving nature and generous spirit. And nothing defined that quality more than observing her become a devoted wife and mother to her two sons, Maged and Nader, both of whom benefitted from her wisdom and counsel.

For reasons I still celebrate to this day, she was forever loyal and affectionate to me. We always had a connection one feels, eldest son and brother/protector for a baby sister. Especially since I had already seen three of my sisters die untimely deaths before Fayza came along, I felt a certain paternal commitment to see her treated well. As long as I could do so, I made a point of seeing to her wellbeing. Thinking America might be good for my sister, I was able to get her over to New Hampshire for a visit, but Fayza was always Egyptian in her heart. So she returned to our native land to be a part of that culture.

Even as she did, she entrusted me with the task of looking after her sons—especially to watch over her younger son Nader when he came to America. I did, of course, and always would keep my pledge to her. We had that kind of cross-connection that defined the meaning of "family," the ones you pray for and yet so seldom find.

So it was a point of pride to Fayza when I brought my wife Bonnie, my daughters Niveen and Amy and my son George over to visit her in Cairo—so much so that she literally got up out of her sickbed to cook us all an elaborate meal. Egyptian meals can be quite large and sumptuous at times of celebration. And my sister labored all afternoon to create the dinner of all dinners. Remembering how she looked at the time, realizing her struggle, made it all the more difficult for me.

By then it was 2001 and very close to the time she died. She didn't want me to know she'd been sick, but I had my ways of finding out. As testimony to her inner will, Fayza put on such a front and poured so much flavor into that singular feast that I remember it to this day. It wasn't that it was luscious. (It was.) It was all the love that went into it—creativity laced with the spice of courage and the sweetness of gratitude—there would never be another like it.

Fayza lived to be 55—longer than most expected—long enough to see her sons grow up and start families of their own. She outlived our mother, and also kept her pledge to Aida to look after our brother Maher once Aida had left these earthly bonds. The thing that brought me the most joy about Fayza was that, despite her health issues, she lived see her son Maged grow to become an adult, marry well and have grandchildren she could openly embrace. With so much love to give, no woman could have enjoyed it more.

My brother Maher becomes a story unto itself that is both a tragedy of failed dreams and a testimony to family loyalty and love. Maher was the youngest brother of six. And, as is often the case with the youngest child, he seemed to summarize all the best aspects of the family rolled into one. He was very smart, made excellent grades in school, and was clearly the best athlete in the family.

Maher loved sports of all kinds, and was absolutely fanatical about soccer (what the rest of the world calls "football"). He lived for the sport, played it whenever possible, starred on his school football team and even dreamt of turning professional. Unfortunately, Maher also suffered from a neurological disorder called *epilepsy*. Not that being an epileptic prevents you from functioning in the world—some of our greatest historical figures suffered from epilepsy. Julius Caesar was epileptic. So were Napoleon Bonaparte, Socrates, composer George Gershwin and golf legend Bobby Jones.

Epilepsy was originally known as "the falling sickness" because it often comes on without warning and strikes the victim in a series of convulsions that, if left untended, can lead to suffocation. People who suffer from auto-immune diseases in general occasionally get epilepsy along with them as a kind of cosmic kicker. Or, as in Maher's case, the syndrome arrived on its own. Most epileptics also have to endure chronic vertigo [dizziness] and temporarily lose all consciousness when it strikes. So it is always best to have someone accompany an epileptic who knows how to treat an occurrence once it strikes.

It is often a failing of good athletes that they believe their physical prowess can help them soldier through anything. Being a teenager to boot and therefore "invulnerable," Maher was convinced he could brave this condition on his own. So, he tended to make light of it until it finally proved his undoing. When Maher was about 16, he was on the upper deck on the third floor of the house we had moved into, on Refaat. Although it had become the (now famous) roost and mini-farm where Aida kept chickens and ducks, it was also large enough for a few extracurricular activities, including practicing some sports.

Making use of its ample space, Maher happened to be kicking around his football when he was stricken with a severe bout of vertigo, lost his balance, slipped over the guard-rail and came crashing down three stories to the ground—onto his head and back.

The fall nearly killed him, but even worse for Maher, it left him largely unable to function. From the day of that fateful accident, my little brother deteriorated and could not go back to school. He became depressed and occasionally aphasic. When you have *aphasia*, you lose your ability to communicate at times and have problems speaking, reading and writing. His mobility now severely restricted, Maher also lost any dreams of athletic achievement he might have still harbored. Not surprisingly, he frequently became morbid. This in turn prompted a kind of biochemical chain reaction in him.

Aida took all her children's well being to heart. But she seemed to feel a special sense of responsibility if not guilt over what happened to Maher. As she had been overprotective of most of us, she thought that she might have allowed him too much freedom. And she felt that her permissiveness, especially in view of his unique condition, had come back to haunt her.

In his way, Maher seemed to embody all that had been best and worst about all the children. He was among the brightest…certainly the best athlete. And yet in the end, his tragic accident at the height of his potential, rendered him the ultimate casualty—physically limited, emotionally damaged and unable to function normally—all those demons of denial that shorten one's life.

Looking after Maher with a special kind of tenderness, Aida personally cared for her son until the last day of her life. And right before she made her transition, my mother transferred the burden of caring for Maher entirely to my sister Fayza. Knowing Fazya's giving nature and trusting her integrity as her own, Aida exacted a promise from her. "Take care of your brother, Maher. Watch out for him. It broke more than his body when he fell; it broke his heart."

Of course my sister kept her word and looked after Maher until he too quit this life at the age of 31. It was perhaps not surprising that Maher died about a year after our mother had crossed over; it was as if he had lost his will. With that special strength that only mothers bring, Aida had been his lifeline; the thing that kept him going. That was the way she was with all of us. It's just that, with Maher, it was expressed more directly. So it was in some way understandable when he too said goodbye.

It is a point of irony that still strikes me to this day, my nephew Maged remembers the night in July 1988 when my mother passed away—almost at the exact moment Aida made her transition—that the lights on the entire block went out and did not come back on for hours. In truth, I wasn't surprised when he told me; it seemed only fitting.*

---

* My brother Saied also has his own powerful recollections of Aida's passing, and an energy that carried with him for the rest of his days. In that summer of 1988, Saied had become seriously involved with a Jewish/American English teacher he had met when he'd moved to Denver. Announcing his intentions to marry her, he was on his way home to get Aida's blessing, and as usual our mother was ecstatic to see her son. "I am cooking all your favorite foods," she proudly announced when they talked. And Saied set out, elated, boarding a plane headed for Cairo. Sadly, upon his arrival, he was told that our mother had died the day he left. Upon learning this, Saied became so distraught that he took this as a deadly omen. He refused to touch a bite of the food, cancelled his marriage plans and eventually returned home. Three years later, his nephew Maged (Fayza's son) introduced him to Nahed, the woman he would eventually marry. Today they live in Florida with two grown children in college. Saied remembers: "I know Aida didn't intend it, but she changed my life from that moment—no doubt for the better; always for that."

CHAPTER 10

PARADIGM SHIFT. THE NASSER ERA

At some point in the next few years there was a shift in my father's behavior. He stopped gambling for a while, seemed to grow more responsible in his position with the government and consciously started to bring a more stabilizing influence into the family. Certainly Aida's enduring love made its constant impact; but this time it was different. I had noted positive shifts in his behavior before; but this time they seemed more remarkable. It may have been due to losing two of their children, George and Nadia, in such a short period of time. It may have been the responsibility he felt toward the rest of us as we grew from boys to men. And though I was too young to grasp all of the reasons, looking back at the era in which we lived, I have to believe that at least part of it had to do with the revolutionary changes taking place in Egypt at that time.

Our personal lives seldom turn around because of one catastrophic event or one glorious revelation. It is usually a series of subtle shifts that create a new construct—and you wake up one day and realize that things are somehow different. Then again, there are landmark events, those political paradigm shifts that shake up the social landscape around us.

That paradigm shift began for Egypt and for us in 1952 in the form of the overthrow of the hopelessly corrupt reign of King Farouk I. In a military coup engineered by (then Colonel) Gamal Abdel Nasser the Egyptian Monarchy came to an end, and the first ever Republic of Egypt was declared. It was a Republic in name only because a few years after he came into de facto power, Nasser removed all other political parties except one—his own National Union Party. Part of this was a countermeasure against an attempt by the Muslim Brotherhood to assassinate Nasser for any number of reasons—not the least of which was the fact that for the first time in Egyptian history, women were somewhat enfranchised. Almost immediately after taking power in 1954, Nasser granted women universal suffrage (the vote), eliminated gender-based discrimination, initiated protection for all females in the Egyptian workplace, and opened the door to them for all aspects of higher education.*

Nasser also instituted enormous agrarian reform, expanded business opportunities and enhanced social services to benefit a much broader base of people. More important was the fact that, for the first time in modern history, there had emerged an Egypt for Egyptians. Up to that point, the Fuad Monarchy had been little more than a puppet government manipulated by the British and French. Nothing bore evidence to this more than their control of the Suez through the (jointly owned) Suez Canal Company. In fact British troops had a base near Port Said and were such a presence in Egypt for so long that one almost came to take them for granted as a permanent part of the landscape.

That changed in 1956 when Nasser had the temerity to seize the Suez Canal Company, throw out the Brits, and return control of the Red Sea waterways to Egypt. This was a political and economic body blow to the British who had actually controlled the Canal since the late 19th Century. Given the stakes involved, they reacted as if their very nation had been attacked. Almost immediately, both the British and the French joined forces with the Israelis to instigate a massive troop movement to retake "their Canal." In a few days, the

---

* The Muslim Brotherhood was founded in Egypt in 1928. Originally an Egyptian nationalist movement, it was eventually taken over by radical fundamentalists obsessed with having a caliphate that would govern all the nations of Islam. As such, they looked upon all moderate Muslims as their enemy, especially when it came to modernizing of the roles of women in society. By enfranchising women, Gamal Abdel Nasser had not only doubled the effective Egyptian power base, he had also committed the worst kind of treason and sabotaged the tenets of Sharia Law. So they set up a plot to assassinate their President in 1956 that was immediately sniffed out and put down.

world looked on at what came to be known as the Suez Crisis, verging on a an impending armed conflict that Egypt on its own was doomed to lose.

Then, overnight, a surprising development took place that shocked the world. United States President Dwight D. Eisenhower supported Nasser's nationalization of the Suez, condemned the tripartite invasion and demanded that France, England and Israel withdraw their troops. After that, he insisted upon a United Nations peacekeeping force and a resolution to the conflict—one that ultimately validated Egypt's right to its own zones of passage.

Shortly thereafter Nasser adopted a foreign policy of "Neutralism" that, considering this final extrication from our British overlords, meant a visible shift toward commercial and diplomatic rapport with the Soviet Union. In the Cold War Era, this did not set well with the US who felt betrayed after their support. But Egypt, during that time, had adopted its own brand of socialism and seemed more intent upon reforming itself internally than worrying about foreign alliances.

Socialism can be robust in its first couple of years of institution, especially when it replaces a decadent one-class society. It opens the veins of upward mobility for the young. And bureaucratic expansion automatically makes government the number one employer across the board. So soon enough 90% of the people were working for the Egyptian government.

Since Karam Iskandar already held a government post, and since practically no one gets fired in a bureaucracy, he actually advanced in his position and attained a degree in Accounting. Math was always my father's strong suit. He had always had an excellent command of numbers (I tend to think that prodigious mathematical ability might have also been my father's undoing where his gambling addiction was concerned. Gamblers tend to be very smart; they think they can beat the odds, which is nothing more than arithmetic perverted by optimism.) This time Karam chose to apply his skills in ways that could enhance his career, and eventually became a highly regarded tax expert inside the government accounting offices.

(According to my brother Saied, one innovation Karam created later benefitted any number of small businesses: That was the practice of "auditing yourself" before the government does. That way small businesses and non-profits would have no surprises, and would be ready for anything that came their way.**)

In terms of our personal domestic fortunes, the most dramatic shift for us came in 1954 when I was 15, Fayza was 8 and my brother Saied was 5. That came at the time when we moved from that single room on Harret Gad Sherrif to our more commodious quarters at El Azizi St.

Our new apartment was double the size—two rooms within a four-room apartment duplex—and with such amenities as electricity, water, and sewer. The new housing was more costly, but it was on a more accessible street in a better neighborhood. At the time, we shared that apartment with an old man named Mr. Riad and his much younger wife. The couple never quite seemed to fit, but they managed to make it work, and our brief time in their company was in general a pleasant exchange.

I had entered high school by then, so it helped to have my own special space to study and prepare for classes. Having good grades meant everything, as did a newly discovered need for peer-group acceptance. And it made all the difference for me socially that I was also able to bring a friend around from time to time without having to apologize for my surroundings.

Even better was the fact that this move had also brought us just one block away from my aunt Moufida and my cousin George. Having spent many summers together, George and I were already close friends. During this age of rites of passage—raging hormones and discovering girls—we studied together, played sports together and soon pursued what we delighted to find had become the "opposite sex." Of course, this was a more innocent time, and the social structure among Coptic Christians was still somewhat constrained. We also lacked anything approaching discretionary income or an allowance that amounted to more than a few pennies. Still, we could fantasize and longed for that time when we might actually become financially independent.

** In fact, Karam got to practice what he taught on a number of occasions. Saied recounted to me the many times he and his brothers saw our father auditing the books of St Paul (Non Profit Organization) every a year, often weeks before the annual audit by agents of the Egyptian Government.

After a year or so, Karam Iskandar was earning enough money to enable us to move into our own private apartment for the first time since I was five years old. It was a move into a residence just across the street from the shared four-room quarters on El Arishi Street and was even a notch closer to "Om George" and my cousin.

The El Arishi apartment was a simple but private residence on the first floor (considered prestigious at the time). We lived in three rooms and a small kitchen and—given the fact that we had indoor plumbing, ample access to fresh water, and electricity 24/7—it seemed to my youthful sensibilities to be the lap of luxury. It certainly made studying easier since I could finally have light any time I needed, and that gave me an academic edge I would put to use.

As it was, my grades were always among the best. I hated being second where my studies were concerned. Even then, I was keenly aware that academic achievement was the only sure pathway to a higher education. And I, for one, was bound and determined to move to the top of the ladder.

Living in this expanded space certainly made for a more peaceful environment where my immediate family was concerned. Amenities are good for that; they can fix those lost moments we all have to spend just getting through a day. So at some level my sister, my brothers and I got to enjoy our parents in ways we might never have before. Repairing the ruptures in our general family was entirely another matter.

Ever since we had parted somewhat acrimoniously from the villa in El Zeitoun, the relationship between my father and his sisters had gone into a deep freeze and had never quite thawed. In fact, nine years had gone by and they still had not spoken. That meant my mother was their only direct point of contact, and that seldom happened because Karam was sure to take umbrage if she tried.

Even though my father was entirely in the wrong, his integrity and his manly pride had been assaulted. It didn't help when he learned about Oginee's and Galila's attempts to undercut his marriage by offering sanctuary to Aida if she would just agree to leave him. Learning of this caused Karam to feel both emasculated and betrayed. To his way of thinking, they posed a threat to the sanctity of his family; and nothing could have been more unfair than that.

To this day I'm certain that—had it not been for my mother's insight and delicate sense of strategy—my father and his sisters would have never gotten back together. Over the years Oginee and Galila still held great fondness for Aida and, through her, a genuine affection for me. Still, the estrangement from my father kept them from direct contact, so they used to stay in touch through a third sister, my Aunt Hekmat, also known as *Om Hanna*.

Recognizing this emotional minefield for what it was, Aida decided to use me as a kind of goodwill courier once we became more financially secure. In the beginning her methods and strategies were simple: Once a month, every month, Aida would send me on a mission to deliver my aunts certain hard-to-get commodities such as sugar, flour, kerosene and oil rations. Then she started building back the relationship from there.

During those years, my aunt Oginee used to work as a nanny for an American pilot who flew for TWA (out of Cairo). And my aunt Galila used to do the same for a wealthy Swedish family who used her services for years. They tried to arrange it so that they could take Wednesdays off, using that time to visit their younger sister Eileen who had been bedridden for a couple of years. So upon the preselected *Wednesday* afternoon, I would be dressed up and sent bearing gifts about the time they arrived. Carrying the "goodwill rations" that Aida had loaded me up with, I would run the whole way and dash up several flights of stairs to their apartment on the fifth floor.

My aunts used to look forward my timely visits, and I especially enjoyed their company because they always managed to give me some small change that was worth about a nickel in US currency. Add to that the apple, orange or some other treat my aunts had put together for me, and I always felt more than amply compensated for my efforts.

Looking back on those special afternoons, I recall that Oginee and Galila would ask about my brother, my sister, my mother and everything else—but never my father. (The wounds and retribution went that deep.) Only my aunt Eileen seemed conciliatory. Perhaps it was her infirmity that made her so mellow. But even then she had a thin contact with life. And that was the first realization I had that, the nearer we come to death, the more ego releases its hold on us.

Even though she was the youngest of the sisters, Eileen had also always been the most frail. When she finally fell ill, the doctors—as they seemed unable to do in those days—could not come up with a definitive diagnosis. And yet all her symptoms pointed to a type of cancer akin to *leukemia*, of which there are about 20 different variations.

Since Eileen had grown increasingly weak over about a 5-year period of time, it bore all the symptoms of *chronic myelogenous leukemia*, which would have required nutrition not medication in order for my aunt to get better. As it was, the survival rate at that time would have been less than 40%, especially since it went unrecognized for decades, and the means used to treat it were inadequate until the 1990s.

Compounding the side effects of Aunt Eileen's illness, no doubt, was her unfortunate choice of husbands. When she was quite young, she married a man named Labib Wahba who was neither very loving nor particularly loyal—nor was he ever sympathetic to her condition.

To aggravate the issue further, Labib had borrowed a substantial amount of money from Oginee and Galila to try to save his sagging textile business. Later, after Eileen had passed away, he declared all family ties severed and refused to pay back their good faith loan. A short time later, he filed for bankruptcy, and my aunts never saw him again.

Through aunt Hekmat, Aida learned about Eileen's passing and managed a rapprochement with her sisters-in-law, using that time to try and mend the tears in the family fabric. At times, it seems as if death becomes a sacrifice that the departed leave for the living to help us put a salve upon our grief and bring us all back together. My father was even permitted to attend Aunt Eileen's memorial service. (He took me with him, I think as a buffer, a safe emotional cushion.) In any case it must have worked; for in a while this rift in the family mended.

After an appropriate passage of time, Aida invited the sisters over to visit us on their days off. Soon enough Galila and Oginee relented, responded warmly to my mother's invitation, and started to show up from time to time. Although Karam was just as stubborn and intractable as his sisters, he seemed quietly relieved at last to have this relationship healed. And to give further evidence of Karam's reformation, he firmly renounced any forays into gambling, swearing off his addiction once and for all…

CHAPTER 11

HOME. GROWN.

Family ties are like the knot at the end of your rope. You never realize how important it is until you get there.

By reuniting with Oginee and Galila we enjoyed not only a healing, but also a turning point in our lives. Seeing the needs of our growing family (and never having children of their own), the two sisters started to pitch in financially, and after a year or so, they purchased a house for us at 19 Refaat St. We moved into this house when I was seventeen; and it is this singular point of reference where many of my best memories reside. This, at last, was home.

This was where I completed my final two years of high school, where I attended college, where I worked, and where I eventually started my married life. In fact, this was the home in which I lived until I migrated to the United States in 1968. My brother Maher was born in this house. It became the ultimate social umbilical for us all—the household Iskandar where there now centered a fragile sense of hope.

The building itself was a relatively large two-story limestone structure—not exactly an architectural landmark, but substantial. The second floor where we lived had five bedrooms, a living room, a kitchen and a bathroom. The first floor consisted of four bedrooms, a living room, and a bath. We rented the entire first floor to various tenants to bring in additional income for my aunts, breaking up some of the rooms into separate apartments.

Living at the house on Refaat proved to be a turning point. This was the most stable environment that our family had ever had. The combination of size, location, and what amounted to a safe neighborhood added to a sense of proprietorship and respectability we had never enjoyed up to that point in our lives.*

In sum, it was a great improvement over anything else we had experienced up to that point; still it was never quite complete and far from perfect. Neither the upper nor lower floors had hot water. In the winter, we had to heat our water and then pour it back into measured portions of cold for things like washing, bathing and cleaning up. In the summer, we all took showers with cold water, which actually came as a relief. There was also no refrigerator, and certainly nothing even resembling air conditioning.

At night we used to see cockroaches crawling or flying around, trying to feed on residue left on uncovered dishes. Sometimes on a summer night, when we dared to leave the doors open for fresh air, we would run across rats and stray cats foraging in our rooms, conspicuously irritated at having their "evening meal" disturbed.

Add to all this the fact that we still struggled financially, especially with my father's occasional lapses back into gambling, and you had a home environment that was paradoxical at best. The good news was that Karam's gambling addiction didn't have the same hold over him that it had when I was a toddler. Still, his lapses of character came often enough. And whenever they did, whenever he came home penitent and broke, both Karam and Aida would try to build back the family coffers by moonlighting. Sewing, tailoring and cutting patterns were bread and butter for our family…and all too often our singular saving grace.

My father was renowned for his talents as a tailor; so there always seemed to be work. But what never ceased to amaze me was the fact that my mother, without so much as a pattern to guide her, was nearly as good. Using nothing more than her hand as a tape measure, Aida was capable of forming out the most intricate of patterns and very often designed dresses

---

* At that time in Egypt if you owned a home or real estate of any kind, you didn't have to pay annual property taxes. And the cost of utilities such as water, sewage, gas and electricity were extremely reasonable.

purely from memory. She almost never had to make revisions, and was intuitive with her clients, seeming to sixth-sense their perfect measure down to the last detail.

Whatever our difficulties in those days, the beauty of Refaat was that at last we had a safe haven we could call our own. Even on those nights when there was barely enough to eat, we could still embrace this point of reference, a permanent roof over our heads that no one could take away. From that time on and in all the right ways the house on Refaat Street was home, and eventually it became the final haven where my father's sisters retired to spend the last few years of their lives.

I have made mention of the fact that Oginee and Galila were both spinsters. And both had health issues, including complications from their diabetes that became more pervasive as they grew older. Foot neuropathy, loss of mobility and chronic fatigue syndrome were some of the challenges that perennially plagued them—some of which had to be dealt with on a daily basis. (Oginee was crippled at the end; and Galila had grown deaf. Needless to say they needed constant care.)

Given the many kindnesses these women had shown our family all through my childhood years, it seemed only right that we should come to look after them later on. As always seemed to be the case, Aida rose to the occasion, making sure that both Oginee and Galila were cared for until they passed away. Before my aunts died, however, they saw to it that the house was signed over, with two-thirds of the deed going to my father and one-third reverting to my aunt Hekmat. A short time later, we bought out aunt Hekmat's portion of the title, and the house became my father's, free and clear.

When you live in a home for nearly twelve years (as I came to do before I moved to America), you have so many memories. And I feel as if I have perhaps placed too much emphasis on the trials and tribulations my family constantly faced.

There were other times I recall as well—the adventures of growing up, and some of the eccentricities that occasionally get washed out in recollection.

I remember the rooftop at Refaat— a sundeck with a guardrail and platform that eventually became host to several kinds of plants and feeders, a chicken coup, a duck walk and later what amounted to a menagerie of rabbits and other "pets" put up there as a kind of a "green-roof" homestead and mini-farm. This was just getting started when I left for America. Later, according to Saied, this turned into quite an enterprise.

At least part of this mini-farm came into being as a means of overcoming some of the financial challenges our family was forever facing. Enterprising even then, Saied worked with his mother to create three different but related businesses, each of which met with some measure of success.

First, they started a rooftop rabbit farm by bringing in 40 female and 3 male white New Zealand rabbits. Some were used for food while others were sold to a store nearby at Baadie Street. The rest, and there were plenty, stayed on to propagate the species.

Cottage industry Number Two came as a result of Saied's Food Technology studies at Ain Shams University. With the help of Aida, and almost entirely due to her fabulous recipes, Saied started a small "Jam/concentrated Juice/European Cheese Factory" and assembly line that sold its produce to local stores and made a nice profit in the process.

Enterprise Number Three was perhaps the most popular of all. That was a luscious Egyptian delicacy called *Mefatta*. Mefatta is concentrated black honey (molasses) with nuts and spices that our family blended together and prepared with the help of a special secret recipe from our Aunt Moufida *(Om George)*. Originally Saied and Aida started out producing around 50 pounds once a week every Friday. And even from the beginning the batch they made was so rich and delicious that it used to sell out almost as soon as it hit the street (sometimes in less than two hours). I have no doubt, had they managed a greater production capacity, my mother and brother could have set up an entire industry from that singular product. As it was, they generated about 5 £ Egyptian each Friday, nearly as much as our father's government paycheck for the week.**

This rooftop garden-farm-aviary was a favorite haunt of my mother's that ultimately served as a total extension of who she was. In truth, from those days to her last days, Aida never kept her house locked up, because she used to cook daily meals for the poor and hungry. So the word went out that anyone who was truly in need could stop by Aida's kitchen and receive food and drink and human warmth …and never a question asked.

I have more than made mention of the disastrous DNA my family inherited from both sides of the family—the diabetes, the rheumatoid arthritis, cancers of various kinds, the hypoglycemia, vertigo and epilepsy make us seem like *The Journal of a Plague Year*. In truth we also inherited a somewhat amusing set of genetic aberrations to go along with them. I have noted my susceptibility to Type 2 Diabetes that finally hit me hard after age forty. What I failed to mention was a rather unusual syndrome I apparently inherited from my father's side of the family—that for the first twenty years of my life I was a *serial somnambulist*.

Somnambulism is just a fancy term for "sleepwalking." And though this condition is seldom serious, perceptions of its cause and effects have altered over the years. Originally it was believed in behavioral science circles that sleepwalkers had no real motivation to do so, nor did they have a conscious connection to the acts that they performed. What they have come to conclude after decades of study is that sleepwalkers are often high achievers, driven by unresolved issues. So they tend, during their sleep state, to try and finish the business they began while they were awake.

I've always been told that I am "driven." And even when I was very young, I would often get up in the middle of the night and study. Later on, when I was really working to achieve good grades in high school, I would apparently

---

** At the time, the value of the Egyptian pound was on parity with the English Pound Sterling or about $2.45. Due to inflation and the vagaries of international markets, the current going rate is about 7 Egyptian Pounds to the dollar.

get up out of bed while still in my sleep state and go sit at my desk, never really accomplishing things but departing shortly thereafter somehow subconsciously satisfied that I had.

On several occasions at Refaat Street, I apparently rose up while sound asleep, walked out onto the balcony for a few minutes and then shortly thereafter returned back straight back to bed.

One night during my college years, while my parents were up, I had actually come from my room and walked over to the couch, carrying a rather large, heavy pillow that I tried to put on a shelf where we kept a large house radio. The pillow kept falling to the floor, but I wouldn't give up, relentlessly trying to put it up there time and time again. For a few moments, Karam and Aida looked on, watching and laughing; they even asked me what I was doing, not necessarily expecting an answer. To their surprise, I informed them quite formally that I was trying to put my suitcase on the shelf.

After a few minutes, I woke up and realized that I was out of bed and yet still in the middle of my dream. Apparently, I'd been dreaming that I was on the daily train that I used to take to college. The next day, my parents told me about what had happened, still amused but quite concerned about my safety at night.

Even after this incident, I continued to sleepwalk until I was about 22. And suddenly, just as inexplicably as I had begun, I ceased my somnambulistic forays altogether. Apparently it is typical of this phenomenon that it takes place primarily with children and young adults, and eventually just stops all of a sudden, quitting the host without rhyme or reason (but with a sense of comic relief).

CHAPTER 12

HIGHER LEARNING: DEGREE OF DIFFICULTY

Ask most people who have gone through the process of getting an education and they'll tell you the same thing: The degree is the easy part. Connecting with your destiny is another matter entirely; that requires a different set of skills.

Especially during the years from 1955 to 1961, education became my highest priority. Since I made excellent grades even at subjects I hated, it was not initially easy for me to narrow down my choice of career. At that time in the educational cycles of most nations, academically sound students were hard pressed to pick out a profession well before they went to college, and to work toward a degree that would channel them directly into their chosen field. Doctor, lawyer, engineer, geologist, accountant, brain surgeon, physicist, pharmacist, Certified Public Accountant—some profession with long term horizons, the corporate embrace and built-in security—these were the options we were encouraged to consider, these and very few others.

Following the dictates of that paradigm, I should say that during my high school years, I was hoping to be a pharmacist. However, at the final national exam, I did not do well enough to be accepted there. (Pharmacists in Egypt were just one notch below MDs, so their initial entrance exams were difficult to pass.) Since I had never failed at anything, and since I had exceptional grades, I decided to double down on my homework until I found just the perfect course of study for me.

One advantage of the Egyptian educational system is the fact that, for those who had the academic proficiency and a high grade point average, tuitions to institutions of higher learning were free. I started college in 1957 and graduated in 1961. I never failed any class and always maintained an A or B+ GPA.

At the time, there were three main Universities in Egypt that enjoyed prestige in international circles. One was Cairo University. Another was Alexandria University. The third was Ain Shams University (also in Cairo).

When I first arrived, I was admitted into the Ain Shams University Accounting College. Perhaps encouraged by my father to do so, I thought I should at least give it a try. But after about three weeks, I realized that I'd be better off doing just about anything else as long as it had more chemistry. So I immediately set about to change colleges into the College Of Agriculture. Once I was accepted there, I continued until graduation.

In my second year of college, one of the courses I took was animal physiology—a laboratory intensive schedule that required small surgical tools for dissection. Other courses such as chemistry required a white laboratory coat that all students attending labs would have to purchase themselves.

I remember that my family did not have the four or five dollars necessary to buy any of the items on my list. So, my father went to his cousins, Dr. Ramzy and Dr. Botrous and asked if they had any extra sets that they could give me. They were kind enough to supply both (second-hand) items. And I was proud to be able to go to school with all the new tools and accouterment in hand.

My own dedication to my studies began to pay off during my third year in college. I began to garner absolutely top grades in all my courses—so much so that the Egyptian Government actually awarded me what amounted to a stipend of £10 Egyptian per month in recognition of this achievement. It doesn't sound like much considering today's exchanges, but at the time it was a big help to us all especially during my senior year. This money that I earned enabled me to help support our family, and to buy things like textbooks, train passes, clothing, and food.

Even with this additional monthly income, I still used every opportunity to save money by continuing to walk 20-30 minutes every day to the train in order to save about 1 US Penny every day. That may seem like an unnecessary expression of self-denial. So perhaps I should put it in perspective.

In my time, Egyptian college students had a lot of trouble finding summer jobs. As a society, even to this day, Egyptians consider restaurant jobs and manual labor to be shameful and demeaning; so many options for supplementary income—even for students—are locked off out of the gate.

Understanding this and being keenly aware that these young men and women need help, has led me to identify, develop and provide support sources for students. One of my priorities in my career has always been to bring financial aid to those students who have shown a dedication toward improving themselves but who lack any visible means of funding.

In my final two years at Ain Shams I found my destined career choice. While in the School of Agriculture, I came to realize that I intensely disliked farming and would have probably made the worst farmer in the world. (It was also one of the first times that I came to recognize the maxim: that, if you observe a situation with a willingness to learn, a *crisis creates an opportunity.*)

First of all, cotton remained the primary cash crop in Egypt. Egyptian Cotton still comprised about 28% of the world market. Because it did, cotton farmers were pressed into a kind of government-structured co-op (along the Soviet model) where it was very difficult if not impossible for them to make a profit and where farmers were virtually treated like serfs. Add to that the fact that boll weevil infestations were rampant, and the *toxafin* used to treat the plants presented even more of a hazard than the infestations it was supposed to prevent—so much so, that when toxafin was sprayed on the infested cotton crops it killed more than 70% of the plants themselves. As if that weren't enough, when it happened to leak across into the water sources leading to the fields, it killed the fish in the water. (And yet, even in view of this dreadful empirical evidence, this was the kind of "solution" that the international chemical companies were using to brainwash both farmers and Government Departments of Agriculture in developing nations alike, convincing them all the while that this killer spray was the answer.) I just knew in my heart of hearts that there had to be a better alternative, and I was equally convinced it was right there in our hands.

What came to fascinate me was the awareness that this was a problem that could be solved by addressing the components of the soil around it; in the growth medium itself…and that the answer was not some juiced-up super pesticide but some undiscovered element buried inside the nutritional fabric of the land. So I was driven by both passion and instinct to address the intricate chemical balance of the soil itself: Soil, the major source of life on Earth! Soil Science!

Pursuing that special course of study was a great revelation to me. Beginning in Ain Shams with General Sciences numbering about 400 students, this unique branch of science was narrowed down by my junior year to a select class of 30. I was fortunate to be one among the chosen few.

By then the year was 1958, a very auspicious time for me. Not only had I discovered my passion through a thorough study of soil science and the infinite landscape it provided, I also met a friend and colleague with whom I would share a close association to this day.

His name was Hassan Moawad, one of the top students in our class. Although he was a Muslim and I was a Coptic Christian, we shared a mutual sense of life—not only to pursue similar degrees in Soil Studies, but also a common passion to achieve great things and to make a positive difference in the world. So, we soon became excellent examples of friends who knew how to reach across belief systems—Muslim and Christian—without bias as to origin.

I'm not altogether sure we could have enjoyed that kind of friendship had we started out in the Egypt of today. But at that time, we visited one another's homes, embraced one another's families…and even learned to pray together.

Even though Hassan went on to the National Research Center in Cairo and I immigrated to America, we have stayed in touch, and have shared our science, our families and our bond—brothers from another mother—and friends to the end.

Like me, Hassan too achieved a Doctorate and several degrees. Although being a Muslim, his path was more direct.

As I would come to learn, the landscape for Coptic Christians in Egypt was about to change. I would soon experience the soft tyranny and subtle shifts in both religion and politics that would make my next steps to mastery and academic achievement more difficult than I ever imagined possible.

CHAPTER 13

LOVE, MARRIAGE AND DESTINY

*"By all means marry.
If you get a good wife, you'll become happy.
If you get a bad one, you'll become a philosopher."*
— Socrates

There are a few times in a young man's life when he makes a decision that will change his life. When those moments arrive they should receive the respect they deserve, and they should be his decision and no one else's. He should never be pressured into any act that goes against his instincts and long-term goals. Having said that, I also believe that all things turn out for the best. And I agree with what Aldous Huxley once said: "Experience is not what happens to a man. It's what he does with what happens to him."

With that in mind, there are three choices I made after graduating from Ain Shams University in 1961 that affected my life dramatically. One of them had to do with my career decision and seizing an opportunity when I saw it. The second had to do with my choice of a wife. The third literally had to do with something called "the luck of the draw." And if one were dealing with gambler's odds, it would have been a lot like winning the lottery. So let's take a look at these three life-changing events in chronological order…

Coming out of the University with a Bachelor of Science degree in 1961, I was facing what every young man in Nasser's Egypt at that time had sitting in front of him—mandatory military service. According to the Egypt National Conscription Law of 1948, every able-bodied man from 18 to 32 had a two-year military obligation whether he liked it or not. (That could be reduced to one year if you had a college degree. Or you could accept a commission as an officer, if you agreed to a 3-year tour of duty.)

Unlike the US or Great Britain where you could opt into a branch of the service such as the Marine Corps or the Air Force and have at least a shot at an interesting tour of duty, the Egyptian military was a farce of demeaning choices. Draftees were poorly trained, poorly equipped and treated by regular militia as little more than forced labor. To make matters worse, the pay was abysmal. And about the only way you could get out of it was through a physical disability.

Since the primary health requirement for young men at the time was having a pulse, I passed my physical and mental exams with flying colors. Given the fact that I had already lost a couple of years of schooling when I was younger, I was more than ready to get on with my career. So the very notion of this deadly drudge of military service was the last thing I wanted. Still, once I finally learned to accept the hypocrisy of bureaucracy, I showed up ready to "take my medicine" and get the dreaded year over with; the sooner the better. But to my surprise (and to my delight!) the year I had lost as a child was suddenly returned of me as if by Divine intervention.

For some reason never quite explained to us, the recruiters had an amassed an overload of inductees. The Egyptian Army didn't have sufficient accommodations—uniforms and housing—for everyone at the time. So, they randomly selected several names from a long list and simply dismissed them from service. I was one of those selected "out." That meant I could at last go into a career that had some meaning.

Lacking the funds to pursue my course of studies as a Master of Science and just having dodged the "bullet" of military service, I realized that my

Bachelor's Degree in Soil Science pointed in one direction: toward a job with the Ministry of Agriculture.

With 90% of all available careers sitting inside the government, seniority and job security came a great deal into play. Even with a Bachelor of Science degree I was still assigned a low mid-level job as a Sector Official for a remote collective of farms. Poor illiterate people tended these plundered tracts of land. What's more, they were loaded up with pesticides and growth media that they really had no idea about how to use. So it was literally like giving a child a loaded pistol. On top of that, these farms were so far removed from the mainstream of Egypt that the only way for me to reach some of these areas was either by hitchhiking or riding a donkey. (This didn't set well at all with me. I was a city boy from the streets of Cairo. This rural life was something straight out of the 18th Century. And the layers of ignorance and inefficiency were so thick you couldn't cut them with a knife.)

I was put in charge of three local group managers who were uninformed and underequipped to do the jobs set before them. Farm production was catastrophically poor, and the media used to cure the infestations of boll weevils came straight from the pesticide/fertilizer bible of soil depletion. No one would listen to me when I made proposals for change. To make matters worse, I came to realize that I was headed into a dead-end job and would be doomed to remain there unless I did something to change it.

That window of opportunity finally flew open in 1963, presented to me by a young former classmate of mine named Monir Abd El Malek. Monir was a bright young man who was also working on the Soil Science Research Staff at the Ministry of Agriculture. Bright, academically ambitious and well on his way to a Master's Degree, he had been offered a position as a teaching assistant at Ain Shams University (something he kept hidden from our employer for fear of being released before the job officially came open).

At that time, there was a little known escape clause in Egyptian Government Services that allowed for a job swap to take place if both parties agreed to it. Assuming their individual credentials and qualifications were a match, the two only needed to secure the approval of a single supervisor and they could literally exchange careers; it was as simple as that.

Having gotten word from a mutual friend that Monir was about to jump ship to Ain Shams, I approached him about the possibility of invoking this little known exchange program inside the Ministry of Agriculture. Knowing how miserable I was out in "Farmville," and finally securing his new post, Monir agreed to make the switch the month before he left for his teaching job at the University.

Monir Abul El Malek was off to a new career, and I had gotten out of the boondocks and into soil research—where I belonged.

For the next 3½ years I used that soil research job both as an inspiration and a springboard. Even though I had used my degree to get this far, I knew I couldn't grow much farther unless I also managed to secure an MS in Soil Science and all the doors that degree would open. So, from 1962 to 1966, I worked diligently at my job, using Fridays and many evenings to concentrate on my studies and toward my MS degree.

Because we were not financially well healed, and because I was now a young man expected to contribute to the family coffers, I couldn't just drop out of the workforce and go back to school. For me, that meant a double life—working five days a week in the Soil Sciences Department at the Ministry of Agriculture, and using my spare time and weekends to slip back into college. This meant essentially working two jobs.

One of the incongruities of socialist governments is that they very often punish overachievement. There are actually limits to the time government employees are allowed to work each week; and overtime, in most instances, is forbidden. That was the obstacle I confronted whenever I wanted to slip back into government offices late at night or on a Friday afternoon to do extra research for school. Workdays always ended at 2:00 p.m., and you had to punch out promptly. Once you did, you were stopped by security before you could even attempt to get back into the building.

Welcome to one of the traditions of "Third World" lifestyles: something called *baksheesh*. The term is linguistically Persian *(Farsi)* but it means the same thing every language—bribery. Lacking the wicked connotations it has in the west, it is an accepted way of getting something done by bending the rules.

So my way to higher education and advanced research for nearly the next four years was to bribe the security guards to let me back in at night; that was the only certain pathway to higher achievement. (I like to think of it as "incentives.") His pay was low; my needs to achieve were a major priority. I helped him supplement his income; he helped me reach a long-term goal. We both benefitted and beat a stupid restriction in the bargain.

Then there was research. One thing we take for granted nowadays is the simple "point-and-click" access to virtual volumes of source material on the World Wide Web. Internet translates to information. All you have to do is Google something or someone these days and *voila!* In ten minutes, you have mountains of data, genealogies and "blogs" that often used to take us weeks to pull together.

All the more reason, I think, to acknowledge the degree of difficulty of what I was determined to do: study, get my hours, write research papers and create my Masters Thesis (on the relationship between ground water and soil salinity). When 1966 came around, I had spent thousands of hours to achieve my MS in Soil Science—something of which I was extremely proud. I now felt I had a field of science I could sink my teeth into; and I could awaken the sleeping giant of mineral studies, as well as the Ministry of Agriculture, into thinking outside the box. I would create some new ways of approaching soil restoration techniques and how they could help restore the world's resources.

I was making a bit more money by then (an additional 3 Egyptian pounds a month). So the new degree would mean an even better paying job. And though I was still living at home, I was now a grown-up, and was starting to get some pressure to become "a family man."

I was only 25 at the time, but I began to feel that perhaps the advice of family and friends might be points well taken. So, I decided to embark on the quest for a bride. It was a matter of custom to do so. Honor dictated the pace. We were a nation of traditions—about a century behind.

This whole passage in my life will be a little hard to explain to those in modern America, the UK and Europe, especially since the era I'm talking about was the late 1960s—the height of the sexual revolution, Playboy, James Bond, The Beatles, marijuana, mutton chops and the mini skirt.

Then there was the rest of the world—my world, Egypt's world, the world of the Middle East—locked in centuries old social customs and unable or unwilling to budge.

Even up to this Millennium in Egypt and all countries in the Middle East, Muslims and Christians marry differently than they do in the rest of the world. Especially in the 1960s and 1970s, social morés in those countries did not allow dating. Sexual freedom as we have come to know it was nonexistent. There were separate schools for boys and girls that went all the way up through high school. Individual interactions between young men and women had to be taken seriously and were generally supervised with family members and chaperones somewhere in the vicinity at all times.

Needless to say—with the resurgence of Jihad, Sunni and Shia radical fundamentalism and the disenfranchisement of women becoming a global phenomenon—the Muslim religion in the Middle East has doubled down on its approach to marriage and the family. Even now, in most countries where Sharia is the law of the land, Muslim men can marry up to four women at the same time, but women can only marry one man.* The divorce *(talaq)* is also very easy for Muslims. The man can tell his wife three times: "You are divorced, you are divorced, you are divorced." And the woman becomes divorced. The Muslim woman cannot do that. Under Sharia she has no rights; she is chattel.

---

* In many Muslim nations polygamy in the pure sense is no longer practiced, but something called *polygyny* is. In the state of polygyny, a man may marry one wife and also have several "lesser" wives or mistresses who do not have the same property rights and ownership of land as the first wife. This was set up for both economic as well as religious reasons. Because, according to Sharia, all four wives of a man are to be provided with equal amounts of legal stature, financial support, property and inheritance, having several wives of equal rank is an enormous financial burden. And this is simply too complicated for most Islamic societies to enforce. So, it is usually one wife and "many mistresses" (or concubines). It's called a loophole.

At that time, and well into the 1980s Christian faiths such as Eastern Orthodox and Coptic Christians were also very structured and quite strict. Of course, as in all forms of Christian faith, men and woman can only marry once, unless one spouse dies. In the Roman Catholic and Greek Orthodox faiths, divorce is prohibited under any and all circumstances, and annulments may only be enacted by Papal decree.

Among Coptic Christians, there is no divorce except for one of two reasons: 1) Infidelity or 2) Infertility. The church does not allow the adulterer in Church again but does allow the victims to attend mass and receive sacraments. So marriage was considered a lifetime contract; and for many, a life sentence.

In the Egypt of my youth, the most common way for Coptic Christians to meet came in one of three ways: through family members (cousins), through Church, or through professional matchmakers. When the last of those options came into play, the matchmakers would run something of a social profile on the families, establish some similarities and then try to hook-up their eldest sons and daughters first.

When this kind of pairing was dropped into the mix, personal chemistry and emotional rapport seldom came into play. The first priority was to shove the families at each other, get the couple in a room a time or two, and hope things would work out. If the first few encounters weren't just a disaster, the families could take the next steps, come to some financial arrangement and plan a wedding. (Love, if it came at all, would have to come later. That was a perk seldom enjoyed in grab bag marriages such as this.)

This was true in Egypt in the 1960s, and I would soon learn all the vagaries of courtship the hard way. Working through a professional Matchmaker named Locas, I was introduced to a batch of women, while he came forward very quickly with an "offer I couldn't refuse."

In the winter of 1964, Locas arranged for me to meet an attractive young woman named Marcelle Adib Ibrahim. Having just turned 24 years of age Marcelle was working at the new Egyptian TV station, a glamorous job description that paid little. But at least she was a "modern" Egyptian woman (in the

New Egypt of Nasser), actually expressing her independence and, by that very juxtaposition, appealing.

Marcelle's family was moderately rich (middle class), and lived at 4 Shara Kotta ("Cat Street"), Shoubra. Her father, Adib Ibrahim, was a retired engineer. She had three brothers and two sisters, none of whom seemed terribly close. Marcelle was the second oldest child and the eldest daughter. So the match, in terms of seniority and family balance, was appropriate.

The first encounters, though not spectacular, were pleasant enough. And to be sure, Locas the Matchmaker was pushing hard for the marriage contract to go through. (Matchmakers customarily received a "finder's fee" or stipend for brokering any successful marriage. And they only got paid if the arrangement was consummated. So, in their way, these people were no different than real-estate brokers or car salesmen and often behaved accordingly.)

As it progressed, the courtship was seamless if not spectacular. A brief engagement followed. Then in Cairo on April, 17, 1966, Marcelle and I were married, and by force of circumstance moved into my family home on Refaat Street. Major Mistake Number 1.**

Over the years, experience has taught me that there are Four Cardinal Rules of Marriage. Follow them and you might succeed. Disregard them and you are both doomed from the outset.

First, marry for love—deep abiding love and mutual respect. Then and only then do you have a chance. For that reason, avoid arranged marriages where two relative strangers are thrust together because they fit some profile, then roll the dice with the rest of their lives and hope for the best.

Second, marry at the right time and for all the right reasons. Don't let anybody or any set of social conditions push you into anything you are not ready for. Men are too often pressured to do it because it is perceived to be socially respectable at some specified time in their lives. Women are often shoved into

---

** At the time housing was close to impossible to find. So to be fair, we had tried to move into our own apartment, only to find that prior to finished construction the owner was renting the same space to 3 or 4 different people. So in fact we had been scammed, and the only option open at the time was to move in with my family.

a marriage because their "biological clock is ticking." Both reasons are specious and wrong.

Third, make sure that your finances are actually strong enough to support a wife and family. Nothing can kill a relationship more quickly than lack and financial difficulty. So many young couples get married when they can barely afford to pay their own expenses. There exists an unwritten law that should be etched in bronze: financial stability greatly improves harmony in the home.

Fourth, live in a space and a place that you as a couple can share on your own. And whatever you do, especially in the beginning, don't move in with your parents.

For the first couple of years, it is difficult enough for two relative strangers to get used to each other even in the best of times. Mixing all that painful readjustment with a whole stewpot of other personalities is a recipe for disaster.

Since poor Marcelle and I broke all four rules virtually from the moment we said, "I do," ours was a relationship that was doomed before it ever had a chance to get started.

I think part of the reason was her new set of circumstances. She was surrounded by a house full of strangers, including a few "younger" brothers at different stages of development. My sole addiction has always been "workaholism." Since I was determined to make a profound impact with my career and my research, I spent more time at work than I did at home. So, in many ways Marcelle and I never really got to know each other.

All this period of adjustment, especially in the first year, was aggravated by my family's financial difficulties, made worse by the fact that Marcelle came from a family that was well to do. She simply wasn't used to having to cut corners.

Add to that a notion called "propinquity." By definition, *propinquity* means *living in close quarters.* Even though my family occupied the top floor of a two-story home with five bedrooms, there were eight of us, including my four brothers and my sister, Fayza. That left Marcelle and me to spend much of our time in one bedroom, and the run of the rest of the house was always populated with other people.

My wife got on well enough with my mother, my father, my brother Atif (who got on well with everybody) and my sister Fayza. My other brothers were another matter. And her relationship especially with my younger brother Saied (who was just entering puberty) was contentious to say the least. They had run-ins almost daily, often followed by shouting matches that would transfer over to me when I hit the door at night. So many of our nights were spent in argument—conflict and conflict resolution—that peaceful moments were treated like buried treasure…and rarely found.

My chemistry with Marcelle was never the best; and frankly the home environment didn't help in the least. Add to that the fact that, once we married, we had very little contact with Marcelle's family. We seldom if ever visited them, and they seemed to show little or no interest in having any interaction at all with my family. It was as if they had just dropped her off at the door and then disappeared.

As it turned out, 1967 was an eventful year for all of us. In fact, it could be marked as a turning point in all our lives.

My career was taking interesting new turns, and I was beginning to consider new opportunities in other countries, including the USSR. Marcelle became pregnant with our first child—our daughter Niveen. And Egypt had descended into a War.

There is no question. War changes everything, especially if you're on the losing side. And our country had been utterly gutted by the loss. As I believe and always will that things work out for the best, it certainly didn't seem that way at the time. And at this point I realize that it might take some explanation…

CHAPTER 14

THE SIX-DAY WAR: AN ORDEAL OF CHANGE

The year, 1967, was one of the most difficult and demoralizing times for several reasons. In the broader sense, in June of that year, Egypt in particular suffered a humiliating defeat against Israel in what ignominiously came to be known as The Six Day War.

The pro-Israeli Western Press heralded as a bold and powerful move on the part of Israel to defend its borders and proactively protect itself against imminent Arab incursions. The nations of the Middle East saw it as an act of naked aggression by Israel, backed and fortified by the U.S. and Great Britain, the primary outcome of which was the seizure of the mass of the Sinai Peninsula (an area about the size of South Carolina). The real truth lay somewhere inbetween.

As a bit of background, Gamal Abdel Nasser, the liberator of modern Egypt had, over the years become something of a living contradiction. In his first 12 years in power, Nasser had contributed in remarkable ways to the progress of Egypt. But, from the time he took over in 1954 up to the events of the Six-Day War in 1967, it had always been a matter of two steps forward and one step back.

Once having created some phenomenal and progressive social achievements by 1958, Nasser made a disastrous attempt to solidify his power base in the Middle East by merging the government, leadership and Islamic philosophies of Egypt with those of Syria and Yemen to form what came to be known as the United Arab Republic (UAR)—a union from which he was to emerge as the locus of power.

Driven by an increased level of anti-Israeli paranoia along with the fact that such nations as Turkey, Lebanon and even Iraq were antagonistic toward what seemed to be Egypt's new role in the Middle East, Nasser was nonetheless determined to forge this union as a kind of buffer region against his perceived enemies. Unfortunately, for a dozen reasons, not the least of which was Syria's plummeting economy and the increased radicalization of Yemen, the UAR came unraveled inside of three years. So by 1961, Nasser was back on his own with egg on his face and looking down the barrel of a world made paranoid by his schizoid foreign policy.

In 1961, he doubled down on his public programs, some of which were visionary and brilliant countered by others that were a botched attempt to superimpose Soviet style socialism on Egyptian businesses. For the first time ever, Nasser instituted co-educational schooling in Egypt up through and including universities (which meant "integrating" the government run Al-Azhar University, the Egyptian Islamic equivalent of Harvard). A Sunni Muslim himself, Nasser paradoxically force-fed the Sunnis in his government to accept all other forms of Islam—including the Shia, Druze and Sufi Muslims they were pledged to destroy. (In religious terms this was like banging a beehive with a baseball bat, because the only groups Islamic conservatives are less tolerant of than "The Infidel" are other Muslim sects.)

Adapting the Soviet model, Nasser also advanced all medicine to include a national health program, and he indemnified all government coopted farms to eliminate virtual serfdom and guarantee farmers a profit. And yet one of his worst decisions over his regime was the socialization of Egyptian industry to the extent that nearly 58% of all Egyptian businesses were "nationalized," meaning they were now at least 51% state owned and were therefore not in charge of their own destiny.

So, by ratcheting up government takeovers of everything, Egypt was headed for economic collapse when Nasser decided to become aggressively militaristic as a means of artificially building up the economy. This did him in because by

1967 the Egyptian military, once Nasser's pride and joy, had become rife with corruption and incompetence.

The turning point for Nasser came in June 1967 when he arbitrarily expelled the UN Peacekeeping force that had been installed at Suez since 1956 and replaced it with his own Army and Navy. He followed that with an Egyptian troop build up in Sinai at the Israeli border, straining Egyptian/Israeli relations to the point that the Israeli Air Force launched a "preemptive" strike, bombing all Egyptian troop positions and some small towns.

So on the morning of June 5th, the Six Day war officially got under way.

It was called the Six Day War for a reason—because, from the moment it began, the Israelis adopted a military strategy first perfected by their old tormentors in the Third Reich—*Blitzkrieg!* Once under way, Israel invaded the Sinai Peninsula with such lightning fast, perfectly coordinated air and ground strikes that it sent shockwaves throughout the entire world. (It certainly sent shockwaves through our family when Marcelle's older brother Ibrahim, an officer in the Egyptian military, was killed in combat in the Sinai. And nothing can strike you so close to home as the death of a family member.)

Clearly outmatched and losing ground in the Sinai with alarming speed, Nasser decided to counter real events with false reports of victory to Syria and Jordan, thus drawing them into the conflict with the same embarrassing result; a virtual wipeout by the Israelis. The truth is that the Israelis had been ramping up for this conflict since the original Suez Crisis and were ready to pounce if given even the slightest provocation. Once they did, they came out swinging like a barroom brawler throwing punches everywhere—at Egypt, at Syria, at Jordan, at just about anyone else (including a "friendly fire" wipeout of an American intelligence gathering cruiser, the *USS Liberty*, that Israelis insisted was a mistake.)

The short-term result of this was a total Israeli victory followed by their seizure of the Sinai and part of the Golan Heights, and a crushing defeat of Egypt's armed forces. The long-term results were even more severe. It utterly devastated Egypt's economy, prompting a five-year economic depression that crushed Gamal Abdel Nasser's will and ultimately broke his heart.

Despite Israel's open expression of religious tolerance, especially in the occupied Sinai Peninsula it now occupied, anti-Israeli sentiment ran high among Muslims throughout the Middle East, and virtually eliminated any illusions they had about its neutrality.

What's more, the Israeli occupation of one-eighth of Egypt's land mass was like a tumor on the body of our nation, tipping the Nasser coalition even more radically over to the Soviets, looking to the USSR for even more logistical support and political pressure on their behalf.

Still, once our country had suffered such a crushing defeat and geopolitical humiliation, the administration of Gamal Abdel Nasser had unofficially lost all steam. He survived another attempted coup in 1968 and remained in power until a final fatal heart attack took him down in 1970. But by then his last gasps were a matter of reflex, and seemed to come almost as an apology.

From 1967 to 1970 especially, Egypt descended into a long run of economic decline and political upheaval. The Egyptian pound plummeted, followed by a hyperinflation for which none of us were prepared. Businesses reliant upon government subsidies just to stay afloat were soon failing right and left. Government layoffs were rampant. Agricultural programs were closed or cut down, and research in departments such as the Ministry of Agriculture were slashed to the bone…and along with them any hope for promising new career opportunities.

I bring all of this up, because so many events in the aftermath of the Six Day War trickled down to me in so many ways—in my financial situation, my career, my continuing education and my understanding of the new "glass ceiling" for Coptic Christians…called Islam.

Mind you, this wasn't something we noticed overnight. But it was gradually taking hold. Religious intolerance, never before felt in Egypt, was now creeping into the conversation and expressing itself through the slow relentless radicalization of Islam inside the one nation in the Middle East noted for religious tolerance. Coptic Christians, who still represented 22% of the population, suddenly found themselves disenfranchised in ways heretofore unknown. And I would feel the fallout within the year.

Given the added pressure of economic lack and a badly depressed currency, we all got hit with accelerated inflation that meant paying twice as much for basics such as food, clothing, fuel and transportation. Gas rationing took hold again. And adequate housing was even paltrier than ever.

The closing windows of opportunity at my government research position, certainly tended to shake me up—especially since I was also facing the ongoing dilemma of an unhappy wife, a newborn baby girl, and a family environment that had become, to say the least, toxic.

This was not helped in the least by the death of Marcelle's older brother in the War itself, killed in action, as were over 10,000 of Egypt's young men in the course of that brief conflict. What caught my notice at the time was how little it was observed by Marcelle's own family. I didn't know her older brother well at all, but his loss hit my wife quite hard. Add to this the fact that she had just given birth to our daughter Niveen, and the timing could not have been worse.

The loss of her brother Ibrahim and continued isolation from her own family seemed to plunge Marcelle further into a kind of post-partum despair—one which I'm sure I was incapable of grasping in its entirety.

My mother, as usual, was there with aid, comfort and counsel to us both, but mothers can only do so much. Ultimately it is up to the couple to work things out if they can. All I knew was that I was bound and determined to be a good father. I had a lovely wife and a marriage that I still believed could be healed. So I had to take whatever steps were necessary to make our lives better—even if it meant changing jobs, changing careers, or even changing countries.

CHAPTER 15

THE POLITICS OF RELIGION

During the time leading up to and through the Six Day War, I had taken a position at the Ministry of Agriculture in Soil Survey and Land Classification. And though my new Master of Science Degree had provided me with added income and ostensible exposure to new career horizons, the aftermath of the Six Day War had put a damper on those for at least the foreseeable future.

So few viable new career opportunities were available at that time that when one finally came up I had to seize it with all the energy I had.

Due in part to Egypt's ties to the Soviet Union, several new academic fellowships became available with the University of Moscow. One such "exchange scholarship" came into our Division for a PhD at that storied institution—with the stipulation that it would be awarded to someone with an MS Degree in Soil Science.

This was tailor-made for me; I could not have been more delighted. Even though it meant a two or three year commitment on my part, it also entailed travel, venturing into one of the oldest cultures in the world, a scholarship to one of the most respected Universities in Europe, and a real chance at a decent income to provide for my wife and newborn child. Even if I could not bring Marcelle with me right away, I felt as if the transfer would provide us time to regroup and bring our life back in to balance.

To whet my appetite even further, I knew at the time I applied that I was the only person who met every criterion for this assignment. I put in my proposal but was summarily informed by my supervisor, Mr. Ismael Raafat, that the "offer" was no longer on the table, citing that "the department did not need someone with a PhD in this subject."

I was extremely disappointed about losing what I considered to be the perfect fit for me. And I had always assumed that Raafat was biased against me due to my being a Christian. (Later, I would come to find out that he had been ordered to do so by "higher-ups" in the Ministry of Agriculture—that rather than give the job to a Christian, they would rather let this "plum position" fall between the cracks.)

Two months later, I had another rude awakening of a similar nature. It was September of 1967 when I saw an ad in the local paper about a job opening for a staff teaching position at the Ain Shams University Department of Soil Science. Since, this was my Alma Mater and a natural connection right into the science school of study where I had earned my credentials, it had to be a match made in Heaven. Or so I thought.

Very aware of the advantage that this offered as well as the fact that the "teaching" position would provide me a straight shot to study for my PhD, I filled out all my papers and checked around to see if I might be able to use my leverage with some old professors or fellow students with whom I had studied. When I did, however, I received my second nasty surprise in almost as many months.

As I was in the process of following up on my application, I was summarily informed that this ad placed "in a public forum" was only a formality to comply with the law and that the position had already been filled, pending final approval. In fact, it had not been; nor was it filled until sometime later…by a Muslim graduate student—with credentials that nowhere nearly measured up to mine.

I should mention here that, in the bureaucratic scheme of things, there was a direct correlation between academic achievement and income. When I finally attained my MS degree, I received a pay raise of approximately 3 Egyptian pounds per month.* My monthly income before that had been approximately £25 (Egyptian). It was equivalent to someone making around

$520 per month in the US in 2009.** That meant my Master of Science degree helped me to receive a 7 to 8 % pay increase. Promotions within the Egyptian Government were based solely on length of service and had nothing to do with performance.

What I was now having to face down was the ugly reality that one's religious affiliations—especially when it came to reaching the highest levels of both academia and government—were a final determining factor and (if you were a Coptic Christian) a possible kiss of death. In truth, this new "academic" climate was being dictated by Al-Azhar and the equal and opposite reaction to Nasser's ecumenical educational programs. The shift to the utter Islamization of Egypt was now underway, and this was one of the first places where it was showing up.

This was made even more evident to me by a key government official and good friend who reluctantly informed that, if I was going to have any hope at all of getting my PhD, I was going to have to consider the unthinkable: leaving Egypt.

This insight caught me like a body blow and yet woke me up as well. After all, I was now a father and a family man. I had responsibilities—to my wife and baby daughter. The thought of spending my life at a dead-end job was utterly out of the question. I had to do what was best for my family. So, I had to take all steps necessary to make that happen, even if it meant leaving my beloved land.

Willing to explore any and all opportunities, I submitted applications to the embassies of Australia, Argentina, Brazil, USA and Canada. My criteria for transfer were solely those of acceptance; so I came to an agreement with myself that I would go only where I was truly wanted.

In retrospect I have to admit that Australia was my original preference. What a forward thinking, energetic country it was. It enjoyed a solid progres-

---

* At the time [1968] the Egyptian Pound was worth about $2.14 in U.S. dollars.

** I draw the analogy because 2009 was our most recent trip back to Egypt accompanied by something of a culture shock.

sive government, a broad expanding economy, and no serious immigration restrictions when it came to bringing in foreign transfers! I knew my Master of Science Degree would at least provide me some leverage to get into those nations that valued scholastic achievement. So Australia seemed the ideal, especially since I had some friends and colleagues already going there.

Sadly, it didn't turn out that way.

Somewhat to my disappointment I was informed that Australia, though they rated my petition highly, had reached an immigration saturation level that forced them to lock off all submissions for an indefinite period of time. And, given the influx of immigrants for 1968, they would not be able to consider my application for at least another five years.

That meant I would wait for my second choice, whatever it was meant to be. And it came to pass within the month: The United States of America.

To say the least, I could have done worse. (How could I have done better?) As I have always believed there are no accidents in life and that things always work out for the best, this was the confirmation I needed: in spades…

CHAPTER 16

SIDESTEP

In February 1968, I received official approval from the US Embassy to immigrate to the Unites States as a permanent resident. Especially in view of the new Muslim paradigm shift in the Egyptian government, my former advisor Dr. Hassan Hamdy had recommended that I pursue a permanent move to the USA as the most viable career option available at the time. Since I had already earned a Master of Science Degree, I was granted a C-3 Priority Immigration Status, automatically categorizing me as an "asset," a nice thing to have on your passport going in. Still, I didn't jump at the move right away; it was a big decision.

It seems anachronistic in a way, but back in the Egypt of my youth nobody ever left the country—especially when it came to career choices. For better or worse, living there was considered a lifetime commitment. As part of one of the oldest civilizations in the world, Egyptians feel a strong sense of national pride; and I was not indifferent to that emotion.

Then there was the ordeal of change; it can be a terrifying thing. And yet when the alternatives one faces are stagnation and confinement, change becomes the highest motivation in one's life. By 1968, I had reached that point. Knowing that it would take me away from my wife and baby daughter was painful to consider. But America was "The Land of Opportunity." Or was it?

It seems almost ironic to me now to recall that back in 1968 America was trapped in the quagmire of the Vietnam War, civil unrest and highly publicized racial strife. So when I solidified my application, I was being told by "experts" (who had never been anywhere outside of Egypt) that the USA was in the midst of a race war, that there would be blood in the streets of the big cities…and because of my swarthy Hamitic appearance I would be looked upon as "black" and might even be attacked. Of course, in the ultimate contradiction, I was also informed by many that the streets were lined with money and that once I crossed the water I would be handed a big wad of cash just for showing up.

All of this was a myth of course. And somewhere along the way, I came to realize that we in Egypt were locked in a political and economic limbo (as well as a communication void). Our economy was going nowhere, and at this rate neither was I; something had to be done.

The point of critical mass for me had actually come earlier when I saw my professional colleagues Anwar Haleem, Adib Roufael and others happily departing for Australia, full of hope and flush with anticipation. So, I resigned my job, settled my income tax issues and purchased a plane ticket on Sabena (The National Airline of Belgium).

My Sabena flight to the U.S. actually had a couple of extra legs—a little side-trip, and a list of diversions I hadn't anticipated. It had been scheduled to take off from Cairo with an overnight stop in Brussels and a two-day layover. As the plane lifted into the air and arched over the Mediterranean Sea, I found myself for the first time shedding tears. The finality of my decision finally hit me, and I had to ask myself, *Why?* Why did I make this difficult decision? Why, after all, was I leaving my home and my family?

Expatriation out of Egypt was unheard of at that time. What's more, I was the first in my family ever to leave my native land. In a way this was a point of no return, because once I left I somehow knew I would never come back to live in Egypt again. Fully aware that this move to the United States was life changing, and even more aware that it was the most important decision I had ever made, I was equally sure that a whole new world was opening up to me.

The hardest part was leaving my daughter Niveen behind. Fathers can transfer hope to their children that they often have to relinquish in others. I could see in my beautiful baby girl a kind of etheric energy and knew, just knew, that she would grow up to be someone remarkable. That meant I had to do everything in my power to make sure that she was provided with every opportunity. So I vowed to Marcelle and the family that I would work hard, that I would not stop until I had made something special of myself, and that as soon as was humanly possible I would arrange for them to join me.

It has always been an instinct with me to believe this: a man is only as good as his word. If you make a promise keep it; your self-respect requires it. Despite our differences, I believe Marcelle trusted in that as well, and that was some consolation.

On the way over, my plane trip became an adventure in itself—one that was almost a harbinger for my entire journey to America—fraught with misadventures that always turned out well and soon enough became seeds of a great experience.

The Sabena Flight that I had booked was rather a special one in that it offered perks, including a kind of "mini-vacation" over to the United Kingdom that, on the surface, appeared to be every young man's fantasy. Please understand that the year was 1968, and London had become the very definition of *chic*. The Beatles, the Rolling Stones, James Bond, mini-skirts and Jaguars—this was the place to be. England was "swinging." I was a poor street kid from Cairo on a kitchen pass to the world who had never had a vacation in my life.

Our flight arrived in Brussels, and after going through customs, I checked into a hotel room already prepaid by Sabena. Then they shot us over to London the next day for what I was led to believe would be a glorious adventure. (At least I could see the British Museum and lunch at an English Pub.) Well…I was in for the surprise of my life; and not in a good way. All I can say is that you should always to read the fine print on any promotional package; and always learn the passport restrictions of the country to which you are traveling.

I arrived in London safely enough, only to discover that, due to the fact that I was Egyptian, I was required to have a "tourist visa" before I could enter the country. Some nations require Visas from others for a couple of basic reasons: One is that they don't have the best of political relations. The other is that a tourist Visa is like a second passport for which the home country can charge a fee—as an income stream and a time-block to anyone seeking employment.

Frankly, Sabena should have covered all the options. Then again, so should I. As it was, the immigration authorities at Heathrow detained me at my terminal for 3 hours, finally giving me orders in both English and Arabic that, because I was Egyptian and lacked the necessary documents, I would have to go back to Belgium on the next available flight.

Once I arrived back in Brussels, I had the Sabena agents call an old friend of mine, Monir Tanyous. I had known Tanyous from Cairo before he had moved to Europe. He had been living in Brussels for a few years by then and really knew his way around.

In a couple of hours, Monir scooped me up in his car, checked me back in at my hotel, and then whisked me off by train into another town where there was a party in full force with food, wine and lots of pretty people. So, we drank, watched dancers, shared some good memories and created some wonderful new ones. By the time the night was over, I didn't miss London at all. Monier returned me back to my hotel, happy about the way that things had turned out.

The next day I got back on to my final leg of the journey and on to my new life in America— more convinced than ever that things would always work out for the best.

CHAPTER 17

COMING TO AMERICA

Three days after I had departed from Egypt, I flew into JFK airport from Brussels, assuming a change of planes would take place automatically. What I hadn't realized was that New York is a very big place with not one but three airports—JFK, LaGuardia, and Newark—and that by changing airlines I was also changing points of departure. After clearing immigration and customs, I was notified that in order to get to my ultimate destination in Milwaukee, Wisconsin, I had to make the switch to LaGuardia Airport. This was extremely difficult because I had two very large, very heavy suitcases, a handbag and a briefcase with my papers, my passport and everything else that seemed to matter.

Then there was the language barrier. It seems inconceivable to me now, but when I first came to America, I barely spoke English. On a scale of 1 to 10—10 being fluent—I was probably no better than a 4.

My first exposure to the English language had been through a Baptist Minister named Norah Gunther who visited my aunts Oginee and Galila from time to time. As memoirs are often like picture puzzles where you find pieces that fit at the most unexpected moments, I have to remember that my aunts Oginee and Galila were also deeply religious women. By original vocation, both began as nuns. They remained missionaries of sorts. And whatever else they came to do later in their lives, they always wove their Coptic faith into their daily undertakings. Part of this spiritual lifestyle included using their home at Shoubra as a kind of a Christian Hostel and Mission—a safe haven where travelers could rest and people could meet and pray every Wednesday evening.

A woman named Nora Gunther knew my Aunts Galila and Oginee from their prayer group and occasionally came to visit them in their home, finally receiving support from them for her Church in Columbus, Ohio.*

A lovely woman, and kind, Nora was the first person who took an active interest in teaching me English. For better or worse, it was through Nora Gunther that I experienced my first immersion in the English language, and it came through reading the King James Version of *The Bible*. Even though it was an earlier form of 17th century English and the vernacular was both poetic and stilted, it nonetheless provided me with some basic points of reference.

(Later on I would go through a 90-day crash course in English at the American Friends Institute, but I'm pretty sure my retention level was less than 30%. So I would occasionally find myself in the midst of several people whose native tongue was English. And it came in such a flurry of conversation that I often felt as if I'd been caught in tornado of words.)

It was also through Nora Gunther that I eventually came to meet an American minister/preacher named Earl Parchia. Reverend Earl was a lovely black gentleman from Milwaukee, Wisconsin who offered to host me should I ever decide to come to America. People often make elaborate promises when meeting in foreign countries—intimations of visitation and the cultural embrace. Some keep their word; others sadly do not. Fortunately for me, Earl Parchia turned out to be a man of honor. So, I decided to take him up on his invitation and made my travel arrangements accordingly.

---

* Apparently after I left Egypt for America, Nora Gunther proved to be as picaresque as she was kind. In fact, after a brief time, she took up residence in one room of our Refaat Street house, teaching and attending regular prayer groups each Tuesday in at the gathering of St Paul Coptic members. Apparently Karam sensed there was a real danger here. And upon his very strong insistence, Nora left Egypt for the USA just a few days before the 1973 War. As it turned out, his instincts were dead right. A few days after the war commenced, Egyptian police came by looking to arrest her (though they never clearly stated their reasons for doing so).

To this day, I still don't know how I ever made the airport switch from JFK to La Guardia. In a blur of interaction and exchanging information I essentially didn't understand, a pair of agents managed to get my bags and get me into a taxi. Somehow through a flurry of expedient check-ins and very helpful people I was sent on my way, arriving at LaGuardia just in time to catch my next flight—Northwest Airlines to Milwaukee, Wisconsin.

As he promised in our correspondence, Reverend Parchia met me at the Airport in Milwaukee, cordially escorting me to his home where I was soon to spend my first few days in this country. The Reverend's wife and children welcomed me warmly. And for the next two weeks I was there, they proved to be gracious hosts who made me feel a welcome part of everything they did. (When he had been in Cairo, I had taken Reverend Parchia to see the Pyramids of Giza, the River Nile and all the ruins as well. So he seemed intent upon repaying the hospitality and took some pains to do so.)

When I first met Reverend Parchia in Egypt I remembered him telling me that he was among "the poorest preachers in America." So you can imagine my surprise when he drove us back from General Mitchell International Airport in a brand new 1968 Cadillac Brougham. And I was even more taken aback when we pulled up in front of his "humble home," a two story 12 room mansion in one of the best neighborhoods in town. I also found out a short time later that he owned some apartment buildings smack in the middle of downtown Milwaukee in what can only be described as prime urban real estate.

I remember thinking at the time, *Wow! This is a Baptist preacher who, by his own description is one of the poorest men in his profession. Good Heavens! If this man is poor, I've hit the jackpot!* Not the most accurate reading on the state of wealth gathering in America, as it turned out. And needless to say I was due for a few more cultural adjustments along the way.

On Saturdays Reverend Earl, his wife, his children and I went to clean the church prior to Sunday's service—a means I suppose of earning my lodging, and a show of good will I was more than willing to make.

I just mentioned being told, prior to coming over, that America was in some kind of racial civil war. Even though there were newspaper headlines about riots in Detroit, Newark and the Watts sector of Los Angeles, generally things were peaceful and there was little of the civil unrest we had heard depicted in such nightmarish terms. Everyone, everywhere seemed quite congenial. And people were innately kind. So whatever kind of culture shock I might be in for, it was buffered by the courtesy of the people in the Midwest.

There were also many cultural differences I witnessed here. For one, I was told that everyone travelled by car, so much so that I was encouraged to learn how to drive before I ever came over. So, when I was back in Cairo I actually took a driving course just so I would be able to get around here. (Good advice, as it turned out. And it served me well in the long run.)

On the second weekend I was with the Reverend and his family I also got a first hand look at The American Way of Death. What I mean by that was the way that funerals and the passing of loved ones were celebrated in this country. I use the word, "celebrated," advisedly because it was both a surprise and a shock for me to bear witness to all this.

When Reverend Earl took me with him to attend a Memorial Service for a departed friend in Chicago, I remember being struck by the fact that this "service" was essentially a celebration. There was a kind of party atmosphere around the lives and memories of those who had passed on, and open displays of such joy and goodwill that it was almost uncomfortable to be around. People were singing and clapping and lavishing the room with an endless run of speeches and recollections. This funeral was borderline opulent, more sumptuous than most weddings in Egypt; and money seemed to be no object.

At first I found it to be offensive, even in bad taste. In Egypt, funerals were somber affairs where everyone wore black, women wept and people wore dark expressions of grief. Once the loved ones were lost to us, we would revisit their memories ritualistically and at specified points in time—after three days, after a week, after forty days, once a year after that. They were always mournful observances, and as such greeted with a great deal of solemnity. Due to the high cost of land, the dead were shoved into shelves underground (much like the old Catacombs of Rome), and their remains were kept in a kind of library of skeletons in some dark caves below the earth.

In America, the Departed all had their own little piece of real-estate set aside at some churchyard—gravesites, markers and, if you were rich enough, your own family crypt replete with fancy carvings.

Then somewhere along the funeral "party" it struck me, we seldom paid such attention our dead in Egypt, certainly not the children who had died; it was like we were ashamed of their deaths, as if they had been somehow an admission of failure on our part.

By this point in my life, I had already lost a brother and two sisters. And to the best of my recollection, their funerals had gone on virtually without note. In a way, it somehow occurred to me that meant their lives had gone by without note, and that didn't seem right either.

It was hard to make much of the loss of my baby sister, Nadia, who was only on this earth two weeks when she suddenly died. It was a "crib death;" that much was certain. And if nothing else, it underscored for me the rude uncertainty of life.

Then again this string of events somehow brought me to remember my brother George who crossed over at the tender age of 14. He was such a good sweet boy, and yet sickly. I remember walking to school with him and watching as he limped from his right side, later almost dragging it along as his condition worsened. Looking back at his incapacity, I realize that it had all the earmarks of MS *(Multiple Sclerosis)*. But as usual his symptoms went unrecognized and certainly hastened his end. Still, the gnawing of guilt we felt over George's loss was profound, and the sense of futility, palpable. We carried it with us, always unspoken, but with a heavy pall in the air.

I felt it now, and though I was saddened by the recollection, it restored my intensity of perception. And at some point I recognized that the American take on life was a phenomenal kind of hybrid where they cherry-picked the best things from other cultures and made them into something special—even the way people looked at death.

Ultimately I had to suspend my biases and recognize that all things must be given perspective, including what I had been tasked to accomplish. After a couple of weeks of staying with Reverend Earl and his family, I realized that I needed to start looking for lodging and a job.

Upon arriving, I had been under the mistaken impression that the Reverend was going to help me find employment. That turned out to be wishful thinking on my part.

I started to look at newspaper ads and got a Milwaukee map. I asked about local buses and bus routes. One day I walked to a job interview and it started raining so hard that my coat soaked through and I was drenched to the skin. I didn't have an umbrella or any kind of proper rain gear, and couldn't lay claim to much of a winter wardrobe. I had come to this country with a lot of unbridled optimism, $223 in my pocket, a lousy command of English, and not much else. Well, as they say, "Reality bites." And it bit pretty hard in a short time.

Every time I applied for a job, the human resources person informed me that I was overqualified for the position—especially when I told them that I held a Masters of Science Degree. From there, I started to apply for lab technician and lab assistant positions with high school and college degree requirements. I was willing to accept any kind of job so that I could begin finding a place of my own and start sending money to my family in Cairo. I wanted desperately to succeed in my new country, but first I had to get work…any kind of work.

The turning point for me came one day when I went for a job interview at a local Milwaukee brewery. The manager who interviewed me was a Jewish gentleman from the Middle East. When he realized that I was Egyptian, he began speaking to me in fluent Arabic. He advised me to go to Madison, Wisconsin where I could find better opportunities.

"It's much more academically open for someone like you," he noted. "And then you have *The University* (common local slang for the University of Wisconsin)!" The man was pretty emphatic in his recommendation; so I took his advice to heart. After couple days I said my grateful goodbyes to Reverend Earl Parchia and his family and hopped on the bus to Madison.

Some things you do in life just have that kind of "good gut feel." And this trip to Madison, Wisconsin felt very much like that kind of decision. Perhaps I had inherited some of my mother's psychic powers after all; only time would tell.

CHAPTER 18

POLITICS 'UNUSUAL'

I checked in YMCA in downtown Madison, paying in advance for a week. Madison was actually a lovely town, essentially designed and city-planned around a pair of lovely lakes—Mendota and Monona. It was the second largest city in Wisconsin, the seat of the State Capital and very much a university town. I use the term, "University Town," because it has a different meaning for me now, one that implies a sense of energy, community intellect and youthful exuberance.

All I knew was the fact that I could feel something of a shift in energy when I arrived. The "Y" was a sparse environment but clean and wholesome enough for my needs at the time. As soon as I could, I began checking the local newspapers for work when an advertisement in bold print immediately caught my eye:

> **Manpower Employment Office:**
> **Come ready for work at 7 a.m.**
> **You will work that day.**

I woke up early the next morning, put on my best clothes and showed up at the address provided in the ad. To make sure they understood my serious intent I made it a point to show up at 6:30 a.m. so I could be first in line. But when I arrived I was very surprised to find that there was no line at all. In fact, I was the only one there…and everyone else was late!

I waited and waited, and no one came until 9:00 a.m. When the office finally opened and a couple of people came in, I rushed to tell them that I had come to work. My English was so poor that when they asked me if I was here to "volunteer," my response was immediate and emphatic.

"Yes!" I replied, with great conviction. "I will be happy to volunteer."

Somewhat surprised by my unbridled enthusiasm, these campaign staffers invited me inside the building, sat me down for a few minutes and told me to wait.

My command of the language was not the best, but I understood the surroundings. These were political or government offices of some kind. And since most of the people I knew in Egypt worked for the government, it only made sense that this was the sort of work that might immediately be made available to me here. Nevertheless, it seemed somehow miraculous that I had managed to find a position like this so quickly; and apparently with little or no competition.

I did think it a bit strange that I wasn't asked to fill out any job application or forms that would ordinarily come with an assignment of this kind. Then again, I assumed, they might be "testing me out," and would get around to the paperwork at the end of the day, once they were able to gauge the degree of my competence. So I was determined to make a go of whatever work my new employers threw my way, and to further impress my co-workers with feats of efficiency and skill.

After a few minutes, one of the staff members took me to a desk and gave me my first job in the United States of America. I was quickly shown how to fold letters into envelopes, stick stamps on the outside and place them all in a very large "Out Box." (Simple enough!) During the whole time I was working I kept thinking how easy this was to do. And I worked straight through until noon, not even stopping for a coffee break or leaving my desk at all.

Finally around midday a woman on the staff stopped by my desk and asked me if I wanted to go to lunch with them. "No thank you," I answered. "I don't eat lunch." (Of course I ate lunch, but I didn't go for at least a couple of reasons: First, I was afraid they would deduct the cost of food from my earnings. I also thought they would deduct the time I took off for lunch from my total day's wages. So, they left me alone there to continue the diligent application of my duties.)

I continued to work until 9:00 p.m. when it became obvious that the offices were getting ready to close. A gentleman who appeared to be the Manager for these particular "government offices" came to thank me for all my hard work and to ask me if I would be able to come back the next day.

I agreed and told him I would be happy to come back to work again the following morning. And, since he seemed pleased enough with the work I'd done, I thought this would be a good time to discuss work hours and wages; specifically how much an hour that I would be getting paid.

I could tell by the embarrassed expression on his face that something was not quite right, especially when he carefully emphasized the point that I had "volunteered," and that volunteers, by definition, worked for free.

A couple of things struck me in that instant: first that I was really going to have to work on my command of English, because until I did I could get myself into a lot of trouble. I also learned that goodwill is the subtext that crosses all barriers of language.

After some back and forth and a linguistic scramble of interpretation, the Office Manager was finally able to get across for me that, by volunteering, I had just spent the last 12 hours working for free, and that my *volunteer* efforts had been for the political campaign committee to re-elect a U.S. Senator from Wisconsin named William Proxmire.

When he saw my apparent shock and shattered response, the man felt so sorry for me that he dug into his own pocket and gave me a couple of dollars. And realizing that I was truly "lost in America" he did something even more special: He made a personal call to Senator Proxmire himself.

Once again, it seems, I had been blessed by fortune. William Proxmire, as it turned out, was not only a US Senator of note, he had recently been rated as one of the Top Ten U.S. Senators in American History. A Democrat who was also a fiscal conservative (a living contradiction nowadays), Proxmire was famous for holding the Annual "Golden Fleece" Awards where he shined the national spotlight on the most egregious examples of pork barrel spending by the United States Government. Later he would work to eliminate them, using satire and humor as his weapons of choice. Surprisingly it worked nearly every time.

William Proxmire was also a true statesman who was highly regarded for taking good care of his people. He actually listened to his constituents and responded to their concerns. And by the sheer proximity of being in Wisconsin, I had become one of "his own." (Of course I didn't know this at the time, but I would learn soon enough. I was about to become a beneficiary of his exceptional social conscience.)

Merely by watching our end of their telephone conversation, I could see that the Office Manager was spending a lot of time explaining my situation to Senator Proxmire. And after listening carefully, the Senator told the man to have me return the next day, and to have someone personally assigned "to run me through the local employment process" and get me some work.

Senator Proxmire was good to his word. By the time I turned up at the Proxmire Campaign Headquarters the next day, they had already designated a young staffer to drive me over to the Social Security Administration, get my ID card and Social Security Number, and then chauffeur me around to various employment offices where I was able to fill out three separate sets of applications that day. One was for the employment office at the University of Wisconsin.

To my surprise, the very next morning at the "Y" I received a message at the front desk to come to an interview for a possible job at the University of Wisconsin, Food Research Institute. I didn't know much about public transport, but I had always been very good with maps. So I spent 45 minutes trudging through the Wisconsin weather to the Institute and even managed to get there a bit early.

It was a small building at the end of the U of W campus surrounded by a cluster of smaller structures that housed a variety of animals, including monkeys and birds. (From what I could tell, there was an energy about the place that I liked; a sense of humanity and progress. Good things were happening here.)

My interview was with the Head of the Institute, a very tall young German named Dr. John Guepfort. Dr. Guepfort was a pleasant fellow. I was surprised at his youth and vitality and even more pleased that he did not seem to mind my poor command of English. On the contrary, rather than be put off by it, he looked over my resume and offered me a position as a Research Specialist for $500 per month.

This kind of money was a fortune to me. (It was 1968!) I accepted the job and started the following Monday.

My life in America had officially begun. I was overjoyed. I had gone from making a little less than what amounted to $80 a month in Egypt to an income that was more than six times as much. Perhaps the streets of America were lined with gold after all.

CHAPTER 19

THE DOCTOR AND THE DISHWASHER

I accepted the Food Institute research job and started the following Monday. During the course of the weekend, I rented a room in a house on Kendall Avenue, close to my new job. As well as being the state capital, Madison was a university community. So the space I found came with the same strange caveats that students so often face.

The rent there was $40 per month, and it was on the second floor of the building. There were three upstairs single units, and the owner was an old lady who clearly had nothing better to do than play the Dragon at the Gate, constantly monitoring the comings and goings of her tenants.

There was no cooking permitted in the apartment—not even a hot plate. My room had a single bed, a peculiar walk-in closet with a window and the only other window in the main room. It was a strange configuration, but I could see its advantages. And I didn't mind the limited space; I was used to small quarters. Besides, the only time I would spend there was to sleep and change my clothes.

At this point in my story I have to confess to an addiction: I am a workaholic. Part of my compulsive work ethic comes from a simple need to eat regular meals. Growing up hungry in a family where "lack" was almost a matter of daily fare, I have always looked upon any personal time as an opportunity to earn supplementary income. With a family back in Egypt, I had to tap every resource. And the only commodity I had to spare was my own sweat equity.

The other part of this comes from something I call *perceptible uses of time*. Frankly, I think most of us waste it. Most people dislike what they do every day and run to the sanctuary of diversion as soon as they get a free moment. I don't fault that social custom. It's just not one I've ever been comfortable with; I've never quite grasped the human urge to be idle.

I found a part time job as a dishwasher in a restaurant nearby downtown Madison. This required me to work evenings and weekends, so it didn't interfere with my regular research schedule at the Food institute. My dishwashing job paid $1.50 an hour, which was basic minimum wage; and my employers pointedly withheld 15 cents out of each hour I worked for a full three months. If I quit before the end of the three months I would lose exactly 10% of the money I had earned. (This was apparently a calculated move on the restaurant's part to try and prevent excessive turnover—due no doubt to the supervisor, a man who seemed to make being tough on the help his personal mission in life. He was passive-aggressively abusive and forced us all to punch out for any break times and then punch in again whenever we came back to work.) I despised losing 15 minutes or 40 cents every day; so there were some days when I didn't take a break.

My food supply for the first week at this rooming house was a large loaf of bread and a package of sliced cheese. I was extremely careful with how I spent the money I was earning. The scant savings I had brought over from Egypt was dwindling pretty quickly. And I had to use my remaining funds to pay for the bus trip from Milwaukee to Madison, one-week's rent at the YMCA, and $40 rent for the first month at my rooming house. All that I had left was about $120, and for me that was cutting it close.

By my second week at the Food Institute, I felt financially secure enough to buy a bag of apples, a dozen of eggs, several cans of peas and beans, and tea bags in addition to more bread and cheese. I found a small plastic bag and some strong string, carefully tucked the eggs and cheese in the bag, and then tied the bag together to hang outside the window, a risky innovation that more or less served as Nature's refrigerator, at least for a while. This was at the end of October 1968 and Wisconsin was always cold. So this remained my default food storage method for about the first six months.

By mid November the winter chill was well upon us, and I was fighting the elements with a wardrobe made in Cairo. Noting my dilemma, one of my housemates, John, told me about a Goodwill store on the North Side of Madison where I might find some (affordable) clothes more suited to the weather. One Saturday, early in the morning, I walked for about an hour to the Goodwill store where I bought an electric teakettle and some used clothes, including a heavy coat and sweater.

While I was walking back, carrying two heavy brown bags, I stopped in the street to take a break. Feeling the need to lighten my load, I dug into the bags and put on some of the clothes that I had bought. When I look back on it, it was a rather odd thing to do, especially in the dead of winter in the middle of the road. So I'm sure I was coming off as eccentric to say the least.

Almost immediately, a police car pulled up alongside me, and an officer got out to ask me what I was doing, who I was, and where I was going. I explained everything to him and showed him my identification. I think the combination of my broken English, my passport date, my nerves and my innocent bungling tapped into his humanity in some way. So, for some reason that surprised us both, he let me go.

Needless to say, after that strange encounter, I made it a practice to dash straight back to the safety of my room. From then on however, I used that electric teakettle to make tea, boil eggs, and warm my peas and beans. It served as my singular kitchen appliance for the rest of my time at Kendall, so the whole experience proved to be well worth the effort.

After two months, I found another dishwashing job much closer to my room and to my daytime work the Food Research Institute. In general my night job was an improvement in every way. In the first place, it paid more money. Second, and perhaps even more important, it was actually a pleasant place to work. It was an Italian restaurant and pizza place called Lombardino, a popular hangout with the university crowd, and one with a much more agreeable management style.

Since I didn't want to lose my original hours, I waited until I had completed the entire three months at my first job, whereupon I collected the [held-back] 10% and happily gave my notice. The manager John cynically noted that he expected as much but nonetheless accepted my resignation, still unaware that he was the reason everyone had left.

The job at *Lombardino Ristorante* was a welcome change. The atmosphere was homey. The food was good. The portions were generous. And the energy and camaraderie of the staff had more of a sense of family than a job.

I continued to make progress at the Wisconsin University Food Institute and truly enjoyed both the working environment and my associates. In my first few weeks, I befriended a professor there named Dr. Dean Cliver who invited me to his house for my first Thanksgiving dinner.

Of course Thanksgiving is a holiday unique to North America. Canadians celebrate it somewhat simply in October with a nice meal on the second Monday of the month. Thanksgiving in the USA always took place on fourth Thursday in November. And it long ago became the singular national holiday that celebrated itself for an entire weekend, especially since it preceded "Black Friday," and the kickoff for the coming Christmas Season.

Since I knew nothing about the Holiday or the customs that accompanied it, especially the merchants' momentum that ran it straight into Christmas, I was both dazzled and confused at this atmosphere of perennial celebration. It was like someone just opened the door to plenty.

Even in my first few days in this country, I came to be constantly amazed at the innate generosity of the American people; and nothing epitomized that big-heartedness better than Dr. Cliver and his wife. It surprised me even more when I found out that Dean Cliver, a white man, was married to a black woman. I had heard so much about the "racial divide" in America that I had not realized interracial marriage was even allowed. I also learned that what I had been told about the problems between whites and non-whites did not seem to exist; certainly not in the horrific ways portrayed to me back in Egypt.

There was such warmth and humanity in the household and such elaborate feasting that I was given enough food when I left to feed me for a week.

What's more, the America I saw on a day-to-day basis was a nation that churned out more abundance that it could consume. The trash bins seemed to be crammed with things that were still quite usable—foods to be eaten, wearable clothes and appliances that still worked.

I found this out for myself on one snowy night in December. I was walking home from my moonlighting job when I came upon a huge pile of castoffs sitting on the sidewalk in front of someone's house. It was seemingly set out for the trash collectors, but there was so much attractive "stuff" in this tangle of hardware and clothes that I felt compelled to sift through it. Sure enough, after I had waded around in this mindboggling mound of things, I came upon an old freestanding television set.

Well, as the saying goes: "One man's trash is another man's treasure." And for me at least, coming across this castoff TV was like discovering gold. (I remember back in 1963 having a large portable transistor radio in Egypt that it took me a month's salary to save up for. And here they were throwing away television sets! What kind of country was this?)

I knew, if I could get this to work at all, that it would enrich my life. I could learn English and have some viable entertainment to boot. I also need to emphasize that these were not the light flat-screen TVs we have all become accustomed to in the last ten years. This was a "box" that had to weigh 70 pounds at least—70 pounds of unwieldy wood and wires that came up to my waist.

Since it was only about two blocks away from where I lived, I decided to carry the TV by myself, hoping to repair it once I got home. I walked with it for a few feet, rested, and then walked again. The most difficult time was entering the house and carrying it up the steps without waking the old landlady. Somehow I managed to summon the strength and succeeded in getting the TV undetected up to my apartment. I was so tired that night that I slept without even trying the television that I had worked so hard to bring back. At that time I was using my shoe as a pillow because that was all I had. Pillows were expensive, and I could put the money to better use.

The next day I tried the TV and, after fiddling about a bit, I was utterly surprised to discover that it worked. It was black and white and boasted a 15-inch screen. But at the time, it was the biggest media marvel I had ever owned. I watched the first Moon-landing on this TV. ("One small step for man"…one giant media device for Alex.) It also helped me with my language skills. Cable didn't exist as such, and the TV had no antenna. That meant the picture would never be clear, but it was free and it worked!

Working at Lombardino certainly brought its share of perks. There were quite a few nights when I was able to take home leftover pizza. And I always looked forward to Sundays because sometimes they would cook too much chicken, and the owner used to let me take at least half the leftovers home with me.

For the first part of every week from then on I lived on that leftover food and saved my entire check from the university. Just before I was able to bring my wife and daughter over from Egypt, I took the plunge and purchased my first car, a small F-88 Oldsmobile. I paid about $300 for it—a cash transaction—and I had the pink slip. Gas was cheap then, only about 22 cents a gallon. Chicken was 22 cents a pound. But the price of a phone call to Egypt was still much more than I could afford.

In spite of all this, I remember those days as some of the best of my life. My star was rising. My hopes were high. My family was coming to America.

CHAPTER 20

MENTORS AND TIPPING POINTS

I have to admit to a method in my madness. Coming to Madison Wisconsin and the University there was also out of design. I was intent upon achieving a PhD in Soil Science. And Wisconsin University was one of the three or four institutions of higher learning where the research in this area was being advanced exponentially over any place else.

Part of the reason for the spotlight on this little known branch of science was the work of Dr. Marion L. Jackson who was a professor in residence at the U of W. One of the most highly regarded figures in the small universe of Soil Sciences M. L. Jackson was, at the time I came to Madison, the President of the Soil Science Society of America.

Determined to commence my doctorate studies under such a notable scientist as Dr. Jackson, I was still doubling down on my workload, first at my research position in the Food Institute, and second at just about anything else I could find. Of course I still had my dishwashing job at Lombardino, but I didn't stop there. I also haunted the work postings on the Bulletin Board at the University of Wisconsin Student Center. Picking up on every possible ad-hoc opportunity I could, I mowed lawns, worked as a night security guard, a "driver's-ed" instructor and (of course) a dishwasher—no job was too small, no assignment too far out of reach. I was determined to get my career under way, and nothing was going to stop me from achieving my goals.

Then I got lucky. A billionaire oilman named Sid Richardson (founder of the Bass Brother's Empire) once said, "Good luck happens to industrious people."

I don't know whether or not that is true. What I have come to understand is that energy is a magnet. And I, in my overwhelming desire to study in the Soil Sciences Department of the University of Wisconsin, drew the most remarkable man into my experience: an associate professor from England and a Durham University Scholar, Dr. Keith Syers.

Although a year younger than I, Dr. Syers was already an accomplished laureate in the study of soil science. Having come to Madison to advance his work with Dr. Jackson, Keith Syers was a part-time associate professor perched on the verge of a grant to advance and develop his theories of the *Solubility of Lichen Compounds in Water.*

The synchronous moment for me came in the form of a posted message I saw on the bulletin board at the U of W Student Center: the notice that this particular individual was interviewing for a graduate assistant to participate in his research. What's more, after I interviewed with Dr. Syers, he informed me that my stipend would be $5000 a year for the duration of the study.

Finding such a position would solidify my status in a couple of ways. To begin with, I could afford to start my course studies for my doctorate. Second, I could finally save enough money to send for my wife and child.

I had been in Wisconsin for nearly a year. So that officially classified me as a resident of that state. Since I was now officially a Wisconsin resident, my tuition to the University dropped down to about $300 a semester. (Once I became a graduate student studying under Keith Syers, I received a monthly stipend [tax-free] for my assistant professorship. That enabled me to put aside $100 a month to pay toward my tuition.)

So my goals were starting to reach fruition. This connection with Dr. Syers was all I could hope for and more. For reasons I cannot explain but for which I will always be grateful, Keith Syers regarded me more as a "fellow," a peer and an equal. He saw past my clumsy command of English and into my body of work—the one I had labored so hard to amass up to now. And yet it was ironic that I, of all people, would be retained for such a study. I was from a desert country and had little exposure to *lichens* or what their possible impact could be upon the soil.

I say that my knowledge of lichens was limited, but it would be impossible for me to have been altogether ignorant of them. There are over 20,000 species of lichens in all parts of this planet. They are blends of three different life forms—plants, algae and fungi—and possess a unique nature in that they are mutable botanical hybrids (and some even flower). They do not really compete with other life forms. They thrive in soggy soil or desert sand… even inside of some rock forms. And certain species of lichens provide excellent food sources for almost every grazing animal in the wild. Others are used as atmospheric indicators—excellent for gauging things like air pollution.

Without getting too technical, let's just say that lichens are one of Nature's great mysteries. In the 1960s, scientists were beginning to believe they might have restorative effects on the soil, *if in fact they could be absorbed.* But the science community was split into two schools of thought about them: School 1 (the majority) were convinced that lichens were "insoluble" and of no symbiotic value to the surrounding soil. School 2 (the Keith Syers school) believed that lichens were highly symbiotic, and they were bound and determined to prove it.

Solubility was the benchmark. This was the principal reason for the study. If we could prove that lichens were soluble and could create a synergistic relationship with the soil and plant-life around them, we could influence the production potential for growth media around the world. It certainly didn't take me long to see that. And though I admittedly knew little about it, I possessed that kind of passionate innocence that all scientists must have—an open mind, a hunger to explore and a dogged commitment to result.

I think that shared passion for Soil Science was what Keith Syers saw in me and turned out to be the guiding reason for his taking me on as an associate. It was the cause for a certain professional respect, and it forged the underpinnings for a lasting friendship.

Once Keith let me in on his research, I jumped in with all the enthusiasm I could summon. I spent all my spare time in catching up and delving even further into the data. I can only emphasize once again; we had no Internet. Research took a butterfly's lifetime, and cross-referencing was a nightmare. So when I went to add in my thoughts I poured over the research for weeks, finally handing him my conclusions in a paper that I'd taken great pains to write.

I handed in my research on a Thursday night and by Friday morning had gotten it back on my desk, redlined with so many markings that it looked like a Christmas card. Devastated and feeling that I had failed him, I came to Dr. Syers ready to receive my comeuppance for such sloppy work. To my surprise, he was encouraging, and emphasized the benign but unpleasant truth that most of my mistakes had been linguistic—those of spelling, grammar and syntax. In doing so, Keith pointed out the obvious and yet managed to encourage me as well.

"Don't feel too bad," he said. "When I presented my first paper at Cambridge, I got it back with nearly the same result—ripped to shreds and with so many corrections that I felt like the dumbest man in the world. Yours was actually pretty good. But there are some issues of language. And we're just going to have to address them, the sooner the better."

By doing what he had done, Keith Syers helped me to realize, if I was ever going to make any progress toward achieving my doctorate at the University of Wisconsin, that I was going to have to improve my English. So I immediately started taking crash-courses—written, spoken and literature—whatever it took to get my command of the idiom to the next level. It wasn't the best. (It still has far to go.) But it was good enough to earn me the kind of credibility I needed.

After a time, I had improved my comprehension and facility for expression enough to join Keith Syers on the project we now shared. And nothing could have made me any more proud than finally to have my name appear next to his about a year later on a groundbreaking study:

SOLUBILITY OF LICHEN COMPOUNDS IN WATER; PEDOGENETIC IMPLICATIONS
I.K. ISKANDAR and J.K. SYERS[*]

---

[*] Department of Soil Sciences, University of Wisconsin, Madison, Wisconsin USA.

> The role of lichens in the chemical alteration of rocks and in soil formation was probably exaggerated by many 18th and 19th century naturalists. More recently, the ability of saxicolous lichens… The formation of soluble coloured complexes when lichens or lichen compounds are placed in contact with water suspensions of minerals and rocks suggests that certain lichen compounds *are sufficiently soluble to function as metal complexing agents, etc.*

What Keith Syers had done for me was to overlook my clumsy command of language and focus directly on the findings themselves—ones that validated his initial tests and moved us from theory to axiom. He had validated a significant breakthrough and allowed me to become a part of it.

In the ensuing months and years my personal relationship and professional association with Dr. Keith Syers turned out to be one of the most pivotal in my life. He was that mystical agent of change that only a few of us are privileged to find. And yet I can never adequately express my gratitude for the impact that he made on my life.

Keith was what one has to define as a high life-force human being, someone who changes everyone and everything he touches. Academics and scientists thrive on a sense of collegiality. And Keith was one of those rare human beings who brought it about in so many ways.

His research and counseling were masterful, perhaps only surpassed by social outings with his peers. And there were many rare Friday evenings when all the members of our research team and other science associates would meet, eat, drink and share in those energizing exchanges where great ideas are hatched and dreams are somehow hammered into action. The Greeks had a word for them: *Symposiums*. Events where men got together and hatched great stratagems, tossing out brilliant ideas like smoke for the whole world to take in. It was a favorite activity of great minds such as Socrates and Plato, Aristotle and Alexander; so why at last not us?

I always admired Keith for the way he optimized his time. 'Work hard, play hard,' was always his philosophy. And ultimately I think it shortened his life; he pushed the envelope. I found out recently that Keith had died in Bangkok in 2011, remarried but having suffered a series of heart attacks. (Finally he had one from which did not recover.) When I was finally informed about it, I was saddened but not entirely surprised. Somehow I think he might have understood and even appreciated that kind of sudden exit from this world. It was after all a life well lived. And he had left a legacy—certainly one for me.

No doubt about it, Keith Syers was instrumental in helping me achieve all I was able to do, and even gave me some shortcut pointers toward attaining my PhD.

It is a peculiarity of achieving one's doctorate in that it is accomplished more by reaching milestones than it is by the mere drudge of years invested. As a point in my favor, I had undertaken some intensive studies while I was working on my Masters Degree at Ain Shams. The curriculum was introduced to us in an optional *ad hoc* series by two celebrated scientists from the University of California (Berkeley). At the time, I remember being somewhat daunted by how demanding they were; but they certainly paid massive dividends once I reached Wisconsin. In fact I later found out that, by completing these courses during my MS studies in Cairo, I was allowed to apply all those hours directly toward my PhD at the University of Wisconsin.

Another course requirement for my doctorate was the mastery of a second language. And since there were no hard and fast stipulations about what that language had to be, I was able to use my native Arabic as my "second tongue." So the combination of the hours saved from both the Cal Berkeley courses and the pre-earned linguistics gave me enough milestones to cut nearly a year off my required course of studies.

The final advantage for me had to be one of the shortest and surest rises ever toward receiving a doctorate, helped in no small measure by some of the papers I had co-authored.

By 1972—just two and a half years after I had begun my studies—I had acquired a PhD in Soil Science that would have at least taken an additional eighteen months to achieve. I had made a decent enough living to bring my wife and daughter over from Egypt. And soon I would work with some of the most remarkable men in the field of Soil Science, in America or any place else.

Truly, and for so many reasons, this was the turning point. I had accomplished a lifelong dream. I had met some exceptional human beings and had redefined my purpose in life. Whatever trials I might have going forward, whatever the personal changes, I was certain of this one thing: I had found the right path.

CHAPTER 21

OF MILESTONES AND MARRIAGES

In so many ways—mostly good ones—the years between 1961 and 1972 seemed to fly by on autopilot. My personal and professional relationship with Dr. Keith Syers and the Soil Sciences department at the University of Wisconsin had put everything on a fast track. And I loved it that way.

Inactivity for me had always been little more than a slow way of dying. In this new incarnation, I was studying day and night and working with Dr. Syers on some groundbreaking studies, while still managing to do whatever moonlighting jobs I could to help pay the bills and get my wife and daughter over from Egypt.

By the end of 1969 my family reunion came sooner rather than later. Around the beginning of December, my younger brother Saied wrote to tell me that the situation with Marcelle had become untenable. So he ended his missive alternately pleading with me and insisting to me that I needed to get Marcelle and Niveen over to the U.S. as soon as possible. I didn't question the real cause for this. They had never gotten on well. Even though he was not, by any means the head of the family, his description of the toxic atmosphere was more than enough to convince me. So my concern for my mother's peace of mind and the well being of my baby girl was quite enough to spring me into action.

I had been putting away every cent I had to facilitate just such a transition. I was at last making some decent money from my university stipend. At that time immigration for family members was easy, especially for someone with my Priority C-3 immigration status. So Marcelle and Niveen joined me in Madison in February of 1970—right in the dead of winter.

I had found an apartment at 912 E. Gorham St., Madison, something we rented for a brief time until I undertook my graduate studies in full. Once I became a full-time graduate student we moved to 504 G Eagle Height Apt in Madison, Wisconsin. This was part of a married graduate student apartment complex owned and operated by the University. The living quarters were clean but simple, and yet in their way more pleasant than any we had ever enjoyed in Egypt.

While I had been apart from Marcelle and my daughter I had hoped, even prayed, that time and distance might heal old wounds and perhaps give us a new beginning. Ironically, even though our relationship showed little improvement, Marcelle did become pregnant once again with our son George.

The French Author, Stendhal once observed: "Separation is to love as wind is to fire. It fans the great and extinguishes the small." There had never been enough between Marcelle and me to fan any kind of flame; and the year and a half apart did little to change that. Despite the novelty of the new world and the movable feast that life in America provided, Marcelle never quite seemed to adjust. And it wasn't before long that the differences that had marred our first two years together resurfaced.

In a way this continuing marital difficulty was less of a burden to me, primarily because I was now on something of a mission where my personal objectives were concerned. Finally being able to embrace what I was meant to do and working with such extraordinary men as Dr. Jackson and Dr. Syers, I was driven quite relentlessly toward my own sense of scientific cohesion.

Since there was no real sanctuary for me at home, I was motivated even more by my work, my research and the fellowship I shared with my colleagues in the department. I was also taking on any moonlighting jobs I could; so nothing was off the table—anything I could do to bring-in additional income.

For the time being at least I was consumed with my career, with my field studies, and with the research papers that continued to spring out of them.

Through the gracious mentorship and partnering of Keith Syers, our co-authored study on Lichens provided me with some initial cache in the academic climes inside the Wisconsin University Department of Soil Science and outside of it as well.

As it turned out, the major part of my PhD work with Dr. Syers spun out into a subsequent study to explain "The Role of Lichens in Rock Weathering and Soil Formation and Mercury In Sediments" in the five Dane County Lakes surrounding Madison. And the research I undertook, published in 1972, ended up not only being my PhD thesis, but also providing me with some valuable professional exposure and peer review.

Funded by the State of Wisconsin Department of Natural Resources as a kind of legacy study, this new band of research was to determine the amounts of mercury (Hg) in local lake sediments. Our task was an initiative to uncover the source of mercury in lake sediments as well as the vertical distribution in the "Madison Lakes" surrounding the city and county.

In previous years there had been a huge spike in the Mercury content found in fish native to the local bodies of water. Up to that point there were no viable means to make this determination without losing significant mercury from the samples themselves. What I was finally able to do was find a method for breaking down and isolating the total amounts of mercury in wet sediment (something that had never been done before). By localizing our sediment cores, we were able to use this method to fingerprint the sources as being high concentrations found in raw sewage from industrial runoffs from local factories businesses, as well as from the discharge of wastewater and sewage sludge directly in the lakes.

Without getting too technical, I will answer two questions most of you are asking right now. Why is it so crucial to locate and isolate mercury? And why is it so difficult to fingerprint?

First, Mercury compounds had been used as a cost-effective agent for any number of industrial, commercial and medical applications—everything from limiting slime from paper mills to providing amalgams and fillings in dental and hospital surgeries. The problem is that mercury is both highly toxic and insidious because it is often hard to detect, especially in soil samples where it affects so much of the ecology. Second, if scientists could fingerprint mercury in all its forms, we could either diffuse it, ultimately changing its personality so that it would become benign, or stop it at the source, reducing it to metallic mercury and thereby rendering it inert.

Far easier said than done, and all I had accomplished with this intensive research was to create one small step in the process—albeit an important one (at least my peer reviews said so).

Over those couple of years that involved the study, I was gaining some reputation among the small circle of environmental scientists, not only for credible conclusions but also for "thinking outside the box" where solid applications were concerned. This was great news that also came at a cost. Research is a demanding mistress. Especially in those days, when I would rush from night-job to class to homework and back, there was practically no time for a private life. Winter, spring, summer and fall all blended into one, as did night into day and weekends into the week. Marcelle worked part time in a nursing home, so we did not see each other much at all.

Even the birth of my son George came in the midst of a flurry. And I came to regret that I didn't take more time to celebrate the occasion. By then it was January of 1971. I was on a record pace toward achieving a PhD, and my personal life was a luxury that I almost couldn't afford. I had hoped that my son's birth might somehow draw us closer together. Sadly, it did not.

To her credit, Marcelle tried bringing the kids to the labs on the weekends to spend some time playing on the lawns of the Soil Science Building. But to be frank (and not at all to my credit), I was simply too engaged with my work to pay them all the attention they deserved.

By then, we had moved yet again to another subsidized housing project at 308 Bay View Apartments, Madison. It was a little nicer than the student housing, but somehow seemed like a change for the sake of change—more space for a larger family and a little more breathing room.

By June of 1972, my star was on the rise, followed quickly by a reality check that caught me a bit off guard. I received my PhD in Soil Science. (And proudly note that I completed my courses without any student loans, borrowed funds or personal indebtedness.) My slate for the future was in effect clear. And yet, for a brief time, I ended up scrambling around to look for a job.

Fortunately, I was offered a six-month "project window" that provided me with a cushion to start looking.

Although my friend and mentor Keith Syers had departed by then, Dr. Dennis Keeney had been spearheading an advanced study on mercury and other heavy metals, emphasizing their effect on the soil. "I want you for this project, Alex!" Dr. Keeney told me. "But you have to know up front that I only have money to pay you for about six months."

Sensitive to the opportunity but equally aware that this project had a limited lifespan, I accepted Dennis Keeney's offer. I approached my research and publication in Madison with a sense of what I would have to call constructive urgency. So I would have to excel at what I was doing, and lay some solid groundwork for whatever came next.

I was frankly a bit surprised that some offers hadn't come pouring in; but I shouldn't have been. A soil scientist is not like an MD, a lawyer, an engineer or a CPA. Public awareness of what we do is virtually nonexistent. And even in the world of academics, it is a rare career choice that not many people share. So, once again for a very brief time, uncertainty crept into my personal life. And it was during that period of uncertainty that Marcelle came up with the notion to go back to Egypt and spend some time with her family.

I thought it was a good idea, even though I didn't express it as such. And yet all the warning signs were there that this might be the beginning of the end. It is usually not a good sign when both parties in a relationship greet time spent apart with a sense of relief. I was still scrambling to find a new assignment, and Marcelle made no secret of the fact that she found my lack of job security unsettling, placing a strain on her and the children that she needed to resolve.

About four months after Marcelle had taken the children back to Egypt, I finally had a concrete offer that came up for me—at the University of Wisconsin's Green Bay Campus. It meant a move to Green Bay and yet another change of residence, the fourth in as many years. Still the money was $600 a month ($7200 a year). So, even presented with a one-year contract, I certainly jumped at the chance.

My move to Green Bay and my new research job was a good one at least for a while. I took on a position as a lecturer and research assistant at the University of Wisconsin branch there, acting ex-officio as an associate professor, teaching in the lab and working in lower Green Bay on a special project: the study of heavy metal distribution down into that particular finger of Lake Michigan.

It is often taken for granted that the Great Lakes of North America—Lake Superior, Lake Michigan, Lake Huron, Lake Erie and Lake Ontario—provide 22% of the world's entire fresh water supply. At present, due to a U.S. trade agreement with the Chinese government and with megalithic corporations such as Nestlé, we are seeing literally millions of tons of water siphoned off into tankers and taken away from this continent, never again to be replaced.

This was not a concern back in 1972 when I went to work there for my post-graduate fieldwork and research. What had become an issue was the presence of thermal discharges from local power plants in Northern Wisconsin especially at or around Green Bay industrial areas. At the time, the plants were using coal as a fossil fuel power source. And the EPA had already targeted "coal" at No. 1 on its Top Ten fossil fuel polluters list. So I was brought in to study the effect of these coal-based thermal discharges into Lower Green Bay.

Technically, any time you use water as a cooling mechanism for power plants, the spilloffs will cause it to return to the environment at a higher temperature. Warmer water contains less oxygen, especially at deeper levels where most fish prefer to reside. This heated influx of chemically altered H_2O in turn affects the ecosystem, stunts aquatic plant life and occasionally kills off large schools of fish. Add to that the coal factor (the number one source of toxic water pollution in the United States at the time) and you were facing the ultimate pernicious cocktail—the pouring of heavy metals such as selenium, boron, mercury and even arsenic into the water and surrounding soil.

One point of emphasis that had triggered accelerated research in this area was Richard Nixon's activation in 1972 of something called The Clean Water Act. As an enactment of the newly established Environmental Protection

Agency (EPA), The Clean Water Act [PL 92-500] was about to put the corporations and the states that housed them on notice to clean up what they considered to be dangerous toxins and pollutants in our water…or else face heavy fines and a probable loss of funding.

The flourishing commercial fishing industry that the Great Lakes had enjoyed in the early 20th Century was already on the decline. So they were already undergoing a kind of damage control in the early 1970s. That made Green Bay, Wisconsin with its heavy industry and proximity to that elbow of Lake Michigan the ideal geographic area for the study.

I had come to Green Bay to advance this research, working under the administration of a Dutchman named Dr. James Wiersma. The research itself was stimulating, and Dr. Wiersma was certainly on top of the subject. But the year itself was almost a detour in my career. In a way, I felt out of the mainstream, and for much of that period I was on my own. With Marcelle and the children back in Egypt, I spent at least the first few months lecturing and moonlighting at alternate jobs—but mostly immersed in laboratory work, which now became a kind of safe haven, a pause on the way to embracing my ultimate calling.

By the time I had accepted the Green Bay research assignment, Marcelle had returned with the children. And right in the middle of this project, I would have to come to one of the most crucial decisions in my life.

CHAPTER 22

PARTING: CHILDREN WILL LISTEN

Careful of what you say; children will listen.
Careful of what you do; children will see…and learn.

— *Into the Woods*, Stephen Sondheim

Once I had gotten situated in Green Bay and had plunged deep into my research at the University of Wisconsin School of Environmental Sciences there, Marcelle returned with Niveen and George (who seemed to have sprung up overnight).

She brought something else with her as well—a sense of no longer belonging. Marcelle never quite adapted to America. Where I saw all its infinite potential, she looked upon it as superficial and simply too hectic for her. She never felt as if she quite fit in and more than occasionally expressed a longing for her homeland. By the same token, I think she also realized that this nation—with all its freedoms, its boundless optimism and its mind-boggling technology—provided certain advantages for the children they could never experience back in Egypt. So she was in some way willing to make a go of our unworkable marriage because of it. I think we were also trying to "keep it together" because, in our sense of the marital tradition, we were both still Coptics.

There is no doubt in my mind that, had we still been living in Egypt, Marcelle and I would have stayed married. Coptic Christian couples just didn't divorce; there was too much of a stigma attached to it. And infidelity was punished with virtual ostracism from the church and its community.

America was just the opposite in every sense of the world. It had become a throwaway culture. Nothing lasted here. It was an economy structured on planned obsolescence and a business philosophy of "cutting your losses." Oddly enough, it always seemed to me that they were actually proud of the fact—that shedding traditions had become the unfortunate by-product of a national coming of age. That new national mindset seemed to have locked down during the protest era of the late 1960s and crystallized in the permissive '70s. So it naturally followed that this mentality would filter over into the family unit as well. Even as early as 1972, one out of three marriages were ending in divorce. And women who cut the marital cord were now being looked upon as "liberated."

That had no small bearing on our decision to separate. And Marcelle and I even tried "a trial separation" for a while. But when we would get together even for a day or two, our negative chemistry would take over, and we would end up in a quarrel. Marcelle unfortunately came from a family where shouting and verbal abuse were fairly commonplace; so that eventually became her default position whenever we reached an impasse.

I hated that kind of energy, but what bothered me the most was the realization that those emotional toxins were now beginning to affect the children. Niveen was highly intuitive, and I was crushed to see the negative impact it was making on her. But now even baby George could feel the spillovers from our constant fallings-out and would even start to cry when in proximity to them. (I cried too, any number of times. But I kept the tears to myself. Men in Egypt didn't cry. It simply didn't happen; not in public anyway. In private, was another matter.)

At some point I realized that our children deserved better than to have to grow up in this kind of interpersonal pollution. Frankly so did Marcelle. By the end of 1973, we had agreed to a divorce, and had even put it on a fast track with a matter of both alimony and child support as a part of the package.

We tried to keep the divorce amicable, but that sadly didn't work out very well either. In another example of utter imbalance and an inability to reason out such times as this, Marcelle quit the job she had in order to exact more alimony and child support. (We might have worked something out had we tried, but mistrust is the only beast that survives in times like these. Frankly it cost us both dearly and ruined any propensity I might have shown at being generous.)

There is an old ballad called "Making Whoopee..." where the lyrics examine the downside of the "hot" relationship that begins as a romance and ends in a divorce. And the lyrics go something like this: "He doesn't make much money. Only five thousand per … Some judge who thinks he's funny, says 'You'll give six to her.' "

My post-divorce monthly payments weren't that far off the mark. In his final decision and awards, the judge presiding over our divorce ordered me to pay Marcelle $360 a month. Alimony was $160 and child support was $100 per child. (Understand that these were "pre-tax" payments.) That left me with $240 a month; after taxes it was more like $155—to live on—rent, food and all.

I didn't really mind it as much as I hated being apart from my children. I knew I could make the money somehow. And I doubled-down on my moonlighting to help pay for everything. The rest I considered "motivation" to rise above circumstances—to find enterprising ways to make additional income.

While I went back to Madison, Marcelle and the kids remained in Green Bay. I used to travel back on weekends specifically to see them. Dr. Wiersma and the staff kept my lab at the University open to me—both for research and as a resource for whatever else I needed. I still had the office keys with me for a long time after I moved back to Madison, so I could complete my unfinished research and as a second residence as a virtual sleepover. Somehow I managed to sleep on the bench so I could save on motel costs. (I had slept on floors and lab tables, shoeboxes and even shoes before. And after a while I was beginning to be convinced that I would go into *The Guinness Book of World Records* for the weirdest collection of sleeping places in recorded history.)

Looking back, I often wondered how I managed to get through it. There were so many tearful drives back to Madison after my visits to Green Bay. I could feel the closeness of Niveen and George and would be racked to my soul with the fragile disappointment they felt every time I had to leave.

Nevertheless, I remained convinced that foul family environments are the ultimate social poison. And I was certain that divorce was still the best option for all of us. Better to have some longing in your life than days that are filled with rage.

I also knew that not having a father was a formula for disaster. (It's now a statistic that more delinquent children spring up from fatherless homes than any other single source.) Still, I was determined to make my children a priority, no matter what. I would be there for my kids whenever they needed me, and nothing was going to stop me from making sure that they had all the comforts, education and mentorship that they would ever need.

In some small way, this difficult time provided something of a spark in me—not only to be successful but also to become financially free. I had a family to provide for; that became crucial in my life. No matter what happened from this point on, I would strive even harder. Success and prosperity became my mantra. And soon I would have both.

CHAPTER 23

MADISON: CRITICAL MASS

Cri-ti-cal Mass – *A compilation of elements required to effect change.*

After my year-long research project in Green Bay as well as my recent divorce, the assignment I soon received from Madison toward the end of 1973 seemed very much to be a sign that good things were on the horizon.

Although my friend and mentor Keith Syers had departed by then, my old soil sciences department in Madison had been funded to undertake a new study that included another assignment for me—at $900 a month, nearly twice the money I had been making in Green Bay.

This study involved water quality and the effect of trace elements and sediments as they accumulated in the fish and wildlife. For that study, I had the privilege of working with another brilliant scientist, Philip Helmke. Dr. Helmke was an expert in hydrology and issues of environmental impact but knew little about soil sciences or the influence of trace minerals. I was able to bring my extensive abilities in breaking down and analyzing trace elements, and at the same time build the mathematical models the study would require. So our collaboration was mutually beneficial and created its own unique brand of scientific synergy.

Our part of the study was funded by the Department of Water Resources and the United States Army Corps of Engineers (whose acronym is USACE, more popularly shortened inside the corps with the popular nickname, "ACE").

Once again trace elements—including boron, lead, selenium and mercury were some of the culprits we were hunting for—this time through examinations of aquatic life (fish) in the food chain.

In most studies of this sort, researchers looked for trace elements in food accumulation by examining the flesh of the fish. What we were pursuing was a study of a small species of fish called *sculpin*, but with a different approach. Instead of taking scale samples, we were drawing tissue from their entrails—the liver and kidney—finding more accurate readings for trace elements, thereby enabling us to correct them to attain higher values.

Not really a commercial fish as such, *sculpin* were of principal interest precisely because of their feeding habits. Rather than prey on other fish (the food chain preference of aquatic animals), sculpin were bottom-feeders and scavengers. They chewed up sediment, minuscule crawling creatures called *punta parie* and microscopic life forms at the base of the body of water in question—in this case Lake Superior. So examining their entrails would give us an almost picture-perfect read of the trace elements in the surrounding terrain.

What I learned from Dr. Helmke was the use of a neutron activation technique—for the determination of trace elements in biological samples extracted and set in a reactor to make them radioactive. These in turn measured the radioactivity and presence of other pollutants. The unfortunate aspect to this study was that it became an around-the-clock job, mainly because changing conditions for the table samples were time-sensitive. This required intensive labor at odd hours for someone like me. So very much like a medical intern or an M.D. in residency, I had to be on call 24-hours a day.

As always, I approached my research and publication in Madison with a sense of what I would have to call constructive urgency. So I would have to excel at what I was doing, and lay some solid groundwork for whatever came next.

During the time I was conducting this study in concert with Dr. Helmeke and his staff, and while I was summarizing all the research we had compiled, I was also busy sending out letters of inquiry to everyone I could think of.

Those letters and resumés went out to no less than 100 different universities, corporations and environmental task forces and everything else that was suggested to me as possibilities for employment.

I loved my work. I was convinced that intelligent soil science was the key to unlocking so many challenges we faced in this world—from correcting pollution to saving the ecology to unlocking barriers to our agriculture and how we fed our world. Our research had shown me all this was possible; all this and more was possible…

What I had failed to grasp were a few of the basics—the iron law of economics, the power of global politics and the insatiable need that corporations have to control their world. To make matters even more of a challenge, as an academic I lived in a world of theory. Scientists like to raise all kinds of questions; they don't necessarily want to solve them. The fact that I wanted to do both made me an oddity in my world.

It doesn't surprise me now, but it did then: it wasn't long before I started hearing back from my 100 different inquiries with what amounted to approximately 100 different rejections. (A perfect record.) In something of a quandary about what to do next, I was apprised of a whole new opportunity that had just come up:

The United States Army Corps of Engineers and a subdivision called "Cold Regions Research and Engineering Laboratory" (Acronym, CRREL). They had a staff position available in their Research Division—Earth Sciences Branch…but only one, and that would be at a premium. It was an assignment on a project for Wastewater Management, a pursuit not altogether dissimilar to the one I was currently working on.

"What the hell is CRREL?" I felt propelled to ask. And a little assiduous fact-checking got me the answer I needed.

What I came out to find and soon was that my timing couldn't have been better. The US Army Corps of Engineers had existed virtually since the Civil War but never took on an officially acknowledged "branch status" until 1974. When it did, it was going to embrace about 12 different divisions of which CRREL would become a critical part.

CRREL had already been an official entity since 1961. It was originally a division set up as a Cold War Research, Resource And Survival Branch set up by the department of defense inside the Kennedy Administration.

At the time, it was believed that the Cold War between the USA and the USSR might actually be covertly advanced on frozen ground—the Arctic and the Antarctic—so every available groundbreaking technique was applied against that strategy. As so often happens when wound-up in the wheels of military strategy, most of the research that came out of it included a composite of peacetime benefits—ones with profound cold weather science, climactic and ecological potentials. So CRREL already had a history and a pretty good one at that. Now, since it was being brought under the aegis of the Army Corps of Engineers, this kind of government reorganization was fraught with infinite potential.

I speak of all this in future tense, because all this was forming together, and the initial budget for USACE was going to be about $15 billion. So funding was never going to be a problem. I didn't know much about the US Army Corps of Engineers. But CRREL—with its emphasis on environmental science, heavy metal studies, ecological toxins and permafrost technology—seemed like the perfect fit.

To my surprise, when I originally inquired about CRREL and the position available there, Dr. Helmke, of all people, tried to talk me out of it.

"Oh, way too good of a job for you," he told me. "Don't waste your time."

Taken aback at first, I ended up using the put-down as all the motivation I needed. What the heck, I thought. I had already sent out a hundred letters. What difference would one more make?

To my surprise, I heard back fairly quickly, and the news was good: they were interested! Still uncertain and a bit unsettled, especially in view of what Dr. Helmeke had told me, I started checking around. And after a bit of probing, I got the answer I needed.

Ultimately a very bright student of Dr. Jackson's named Howard May came to provide the missing moral support and encouragement I needed to push ahead. When I mentioned this affiliation as a possible career opportunity, Howard was highly conversant when it came to CRREL—the fact that its lab was quite well known for its many accomplishments, and that it was an agency with powerful affiliations both in the Army Corps of Engineers and the USDA.

Part of the challenge, for me at least, came with the fact that it was in some place called New Hampshire—a small state up in the Northeast— a remote part of America that, as far as I was concerned, may have well been Siberia. But Howard put my concerns to rest, at least for the time being.

"This lab is well known. It's a good place. It's in a town called Hanover, and Dartmouth College is there," Howard was careful to add. (And he was as emphatic as Helmke was cautious that this was the place to be.)

I knew Dartmouth was in something called the Ivy League, which meant it was academically sound. But that was about the extent of it. The rest was a mystery.

Intrigued but still somewhat reticent, I prepared my formal application and submitted it to CRREL, and after a few weeks they got back in touch with me. What I found out to my surprise was that they had received more than 80 applications for that single position and they had narrowed them down to two candidates. I was one of the two. The other applicant was from Ithaca, New York (the home of Cornell, another Ivy League school). And when in doubt companies often tended to go for the "local boy." I knew that much, and didn't think much of my chances. Nevertheless, they considered me seriously enough to bring me in for a series of interviews and a seminar or two for indoctrination.

I didn't hear from CRREL for a while and figured that they probably went in another direction. So I decided at last to inquire, just to get some finality and clear things up. To my surprise, they informed me that mine was the application selected, and that I was to work with a gentleman named Dr. Dwayne Anderson in the Earth Sciences Branch.

CRREL's official offer came in shortly, and I was in a quandary. The money was good, more than double the $18,000 I was making at the

University of Wisconsin, and almost as much as I had made to that point in my life.

Still, it meant that I would have to leave Wisconsin, the only roots I had known here. It also meant leaving my family halfway across the country. Then again, I came to realize that I could provide for them and still have a livable income…and maybe I could start a new life for myself.

After all, I was a bachelor now—free to move around as I pleased. And New Hampshire might just turn out to be where I was meant to be.

CHAPTER 24

CRREL

Scientists are always the first to realize this: Most of us know very little about this planet we inhabit. There are so many things that shape our world of which we are utterly unaware. They affect the air we breathe, the water we drink, and the very earth that moves beneath our feet. They affect our health, our longevity and every aspect of our energy production, our food chain and the infrastructures of our future. And so it is science and scientists that tap into these hidden realms and bring us solutions without anyone having to give it much thought.

That was the world that CRREL lived in; and now I would get to be a part of the universe that the Cold Regions Research and Engineering Laboratory had created: *Biogeochemical processes in Earth Materials (soil), Cold Regions Infrastructure (building permafrost envelopes), Hydrology and Hydraulics, Water Resources Geospatial Applications (in locks, dams and rivers), Environmental Fate and Transport Geochemistry*—all these officious titles and more formed the core of what we were designed to do. And it would probably take a Glossary of Terms to explain. So I won't; not here anyway.

Let me try to simplify all this in a sentence or two: If it had anything to do with removing impurities from the soil, toxins from earth and water, cleaning up pollution by massive engineering excavations through reclamation processes and working on ways to modify climate change, it probably would have involved CRREL. Environmental analysis, reclamation, purification and improvement were our life's purpose. And ICE was our weapon of choice.

That's right, I said *ICE* (and I don't mean Immigration and Customs Enforcement)! Some of our primary areas of focus were frost effects on large-scale ecosystems and soil systems, the study of ice effects in navigable waterways and experimentation with permafrost and frozen ground (especially when it came to controlling toxins).

Of course there is a lot more to it than that. And as scientists working with every aspect of the environment, especially in the ways it involved our Earth's surface—land, soil and waterways, CRREL is similar to a university, in that it consists of engineers and scientists working together to solve large scaled problems as well as performing research.

A lot of our work, especially during what was once called "The Cold War," had to do with information gathering and applying environmental tactics against old adversaries like the Third Reich during World War II and later (in the 1950s and 1960s) the USSR, especially when it came to our competition over the frozen Tundra of the Arctic and Antarctic Regions.

But even more than politics, science makes strange bedfellows. Adversarial relationships often end in a kindredship of spirit, shared information and a pooling of resources—against the wiles of Nature, against the unknown, in the spirit of problem solving and hunting the hidden dragons of discovery.

When I came to Hanover, New Hampshire in the spring of 1975, my assignment was as a research chemist for a new Wastewater Management project to identify and extract toxins and heavy metals from wastewater as a means of converting it back to a viable medium for effective crop productions. A man named Paul Murrmann had done the groundwork on the project. Right in the middle of it, however, he had left CRREL to take on a position with the United States Department of Agriculture (USDA), leaving CRREL virtually high and dry (every pun intended) when he left. So I was the one chosen as the project manager to get the program quickly back on track.

In the field I had chosen—especially soil sciences—scientists and researchers are pretty much at the mercy of university grants or government funding to accomplish our work. Especially in the areas of land management, soil analysis, agriculture and the environment, corporations (who have the funding) are usually on a fast track to advance their own agendas. ConAgra, Archer Daniels-Midland, Cargill, Monsanto and other Agra-chemical giants were already busy advancing their own pesticides-chemical fertilizer-GMO growth cycle agendas. So they weren't terribly interested in funding sane "natural" solutions to creating healthy growth media for our food.

That left government as the last best hope to accomplish this kind of objective. Ironically that made ACE, the USDA and the EPA partners in this pursuit—another agenda initiated by Richard Nixon before he left office. In a way it seems ironic now that Nixon—so vilified over Vietnam and Watergate—did so much when it came to advancing issues such as OSHA, the EPA, clean water and environmental causes. Almost every one of these initiatives bore and impact on my research and my career. And indirectly Richard M. Nixon had also affected my life in yet another way.

In 1972, the year I had finished my PhD, I had finally been granted my US Citizenship, a proud moment that carried with it a kind of danger zone. As an eligible male (and a newly divorced one at that), I would have technically qualified for the military draft that was still in effect. Mandatory military service had been initiated in World War II and kept on officially until 1972 when it was to have a two-year window to be phased out. Technically, I still fit into that window but fortunately was dropped from the rolls due to two changes of policy. One was the fact that, in 1971, the US Congress had pulled the ceiling for draft-eligibility back to the age of 30. And I was already 34. The other was the imminence of closing the draft altogether and going to an all-volunteer army. So my name had originally been placed on a list that had now been cut to the bone.

Up to the point, I had already been blessed by good fortune in so many ways, I had to believe in the favors of fortune and God's perfect plan for me. So by the time I took off for my new job at CRREL, I had never felt freer in my life.

Essentially, I was a bachelor, answerable only to myself. As long as I met my support obligations my time and pursuits were my own. (I had to admit it felt strange for a while; it took some getting used to. Freedom is a joy to have but it comes with a warning label: "Use Me Wisely." So it was something I simply had to relearn.)

When I finally packed up and left Wisconsin for the uncharted waters of New England, I took a week to travel to New Hampshire. It was the first real vacation I had ever had. Eager to look good to my new employer and not wanting to hit them with a lot of overhead going in, I rented a U-Haul, loaded it up myself, and used it to take my books, my bed and a few incidentals halfway across America. The out-of-pocket cost for the Government was just north of $500…

As part of my changeover and new affiliation, I was granted an allowance to lodge for a month in a hotel of my choosing while I looked for a place to live. Again, anxious to do the right thing, I stayed one week at the local Howard Johnson (still a major hotel chain in those days) and assiduously set about looking for a home.

In a week's time I had found small one room house on Greensboro Road in Hanover.** So, after only eight days nearing the tail end of a New England winter, I was ready to hit the ground running at CRREL.

Ironically one of the first things I heard about once I'd arrived was that another scientist the "Lab" had transferred from a university in the South had moved 11 rooms of furniture, a car a boat, a wife, three children, and two horses for a total moving cost of about $35,000.

If working for CRREL in future years had taught me anything, it was these three things: First, money was never an object. Governments, who need to justify themselves at all times, are by nature the very mechanisms of extravagance. They make everything difficult and don't mind paying for the privilege; red tape comes with a lot of funding behind it. This is true of all governments, be they Egyptian or American. (I've never understood squandering anything, but now I've come to understand it as part of business as usual.) Second, if you present your case intelligently, you can get funded for just about project

---

** At the time, I had looked at an apartment at a lovely place called Valley Garden Complex. But at the time I could not afford the rent. Fifteen years later, once I learned the fine art of real estate leveraging, I was able to buy the entire 32-unit development.

you truly believe in. It may take some convincing and good politics. But IQ trumps bureaucracy (much of the time). Finally, the discovery came to me the hard way: the better you get, the more you invite scrutiny. And once you work for the government, you have them in your life in ways you never imagined, especially if they think you're doing something important.

In the coming months and years with CRREL, I would make some groundbreaking discoveries, co-author position papers with some of the most brilliant scientific minds in the world, and be recruited by several U.S. intelligence agencies. I would also come to meet my soul mate and my wife for the rest of my life…and turn into an entrepreneur in the bargain.

Altogether I would come to know that I'd found my destiny. And once you have, the universe unfolds before you.

CHAPTER 25

EARTHBOUND: A PRACTICAL SCIENCE

Changing careers from nearly seven years in academia and the world of scholastic research came as something of a culture shock, and yet a good one. CRREL quickly became a reality check for me. Switching over to the hands-on practical applications of both the Military and the US Government immediately takes you beyond research and into result. You are actually expected to work toward a conclusion that may be put to the test. And there will be challenges, occasionally fierce ones, from opposing points of view. So in this challenging new scientific environment, you had to be ready to defend your positions; and you had to do your homework in every conceivable sense.

Part of the reason had to do with the stakes at hand. Academics have relatively few pressures in that they are asked to present their findings and receive peer review. Academic science is all about theory and discovery. Practical science, especially as it regards economic leverage, is always under stress of a different kind.

I was in the real world now. The US Army Corps of Engineers was expected to come up with solutions and act upon them. CRREL was challenged to provide at least some of those solutions. The wild card in much of this was driven by economics on both sides: and usually it involves a power struggle between chemistry and Nature, not to mention economic forces that drive one against the other.

I was flung into this kind of problem/solution paradox almost from the moment I hit the door at CRREL. When I did, I was immediately led to a laboratory down in the basement of CRREL's three story offices. (So it occurred to me that I might literally and figuratively have to work my way up from the bottom.) I was given an assistant, Dan Leggett, and we were challenged with The Project:

At the time, CRREL was trying to find ways to deal with wastewater without having to resort to channels to dump it off. So they were convinced it could be utilized and recycled provided the toxins and pernicious bacteria germane to the samples they found could be neutralized in some way, returning a natural nitrogen balance to it without overloading the mix. And though the intentions of the study were good, their applications were ill-advised because they were injecting *reagent grade* heavy metals in high concentrations into the wastewater precisely to simulate the level of industrial content. And just as surely as they were killing the bad bacteria, they were killing the good bacteria as well. By the very nature of the intensity of the "media" CRREL was using, they were getting overbalanced numbers, and always would.

Applications of chemical compounds such as copper sulfate, zinc oxide and cadmium were destroying the bacteria, but they were also killing the plants. So, if I may resort to a medical analogy, it was very much like the old cliché: "The operation was a success, but the patient died." In applying this heavy metal reagent grade technique, our predecessors had ignored two cardinal rules of Soil Science: 1) If you can understand what is in the soil, you can control it; 2) You have to apply a mathematical model before you can attempt anything. (They had disregarded both guidelines and the result was a proportionate failure...up to that point.)

It is important to note that when we began this project we had been presented with collected data but had not been given the means to cross-reference it. In those days, your average computer was the size of a large dining room. PCs were unheard of (until 1982). We had computer data cards and some readouts but nothing even resembling an Internet. So research could have taken months to retool, reassemble, and report.

Fortunately, I soon applied a solid mathematical model to the problem and then phased-in a whole new set of engineering criteria and design. Using soil samples from the terrain north of Hanover, we were able to recalibrate the samples and apply gradual mathematical increments to find the solution. The result was an ultimate success. And a written position paper on the experiment followed: *A Study And Report On Urban Waste As A Source of Heavy Metals in Land Treatment.* (This was the first set of findings that I had published on my own.)

On my first project, I had given the right advice and come to conclusions that satisfied both CRREL and ACE, thus bringing me up a notch higher in the world of statistical abstracts.

If I failed to make mention of this before, I emphasize it now: In the world of science and peer review the watchword is, "Publish or perish." When I say that, I mean the rules are simple. For your work to be considered valid, you have to reproduce your findings; that is to say if you mean to expose them to any serious decision makers.

(Just a note: There are more than 1.2 million books published globally every single year. Nearly two thirds of those volumes are dedicated to the findings of serious chemists, physicists, engineers, researchers, biochemists and [yes!] environmental scientists around the world. It is in that realm where so many decisions are made. And it is proof positive that our life's pursuits have meaning. Sometimes they are mere bricks that form the wall; and on rare occasions the entire building. Still, we must produce and we must report. It is that hidden galaxy called Science.)

By coming to the research conclusions I had and by presenting the evidence at my disposal, I began to gain some notice in the world of international science, including nations such as Canada, the USSR and Israel.

A second study we conducted in 1976 (reported in 1977) extended the umbrella of effective Wastewater Management to cover an even broader area of both environmental study and possible conservation.

Please understand that for decades *wastewater* was looked upon as an environmental problem—something dirty and dangerous to be either funneled into a benign recyclable (if limited) state or disposed of altogether. With our studies inside CRREL in 1976, we intended to show that, by working with Nature and available terrain, we could utilize existing soil in some very positive ways: 1) to filter and ultimately purify the wastewater and 2) convert it into a substantial growth medium without the addition of chemical fertilizers or pesticides.

Of course when you present this kind of structure, you fly in the face of two gargantuan interest groups—big Agra and big Chema—and all the megabillions they get from flooding the world with their artificial solutions.

I have always believed that Nature should be the first resort. You work first to facilitate all her earthly rhythms. Failing that, we go to humankind for the solutions; but even then they should try to be the right ones. In this case, we were applying and expanding three basic soil-solution techniques to help channel wastewater into a successful growth medium for healthy crops. These methods were as follows: *overland flow* where wastewater is applied to sloping grassy terrain and allowed to filter through it on the way to nitrifying the native soil; *slow rate systems* where the wastewater is allowed to percolate into vegetated land, and that way act as a natural cultivation medium; *rapid infiltration* where the wastewater was sent through silt and sandy loam that, on its own, acts to quickly purify the water and apply it to the surrounding environment as an agent for recharging the groundwater.

These techniques were not new and had been undertaken in the past. But prior paradigms for application required storage reservoirs for the water to remove them from the ultimate medium as a safety/purification staging area. Our theory was that the terrain itself should be able to act as a natural purification system for the wastewater (without the need for storage or chemical treatment); and this could be accomplished without barriers from the source to the native soil in question. Again, we published these results showing that the surrounding terrain in every case served as a natural filter for the wastewater and, on its own, was capable of turning a potential environmental villain into a nutrient and (therefore) an actual ecological benefit. In 1978, we assembled a corroborative set of findings that we submitted to *The Journal of Environmental Quality* where they were reviewed and published.*

We published these findings and others dealing with Wastewater, leading to a complete body of work, including our research and that of our colleagues, which I was ultimately able, in 1981, to edit and compile into an entire anthology of mathematical models and scientific conclusions entitled *Modeling Wastewater Renovation*. I grant you that it will probably never make the *New York Times* Non-fiction Bestseller List. But, even after 35 years, it is still regarded as the Bible of land treatment where applications of wastewater and effective renovation techniques are concerned.

All in all, it was the beginning of 10 books and about 80 professional papers I have been privileged to have a hand in creating over the years—dealing with everything from recycling natural resources to frozen ground technologies that may be used to forefend Nuclear spilloffs.

At this point, I also acknowledge that scientific research is almost never done in a void. Scientists must, by nature, be team players. We have to be both open-minded and humbled by the fact that the more we discover, the more we realize that the process of discovery itself is endless, infinite and daunting. For that reason, we invariably pool our resources. And when we report our findings we are all parents of the result; we are all authors of the papers we produce.

If I have stood out as an author or an editor, I suppose it's because I've been willing to do the dirty work of gathering, compiling, editing and defining all the studies these amazing people have done. That has been both my burden and my honor to be a part of. That's why you will always see the names of my collaborators—from Keith Syers to James Wiersma to my future collaborator Magdi Selim—my valued colleague, "co-conspirator," and friend on so many science papers, abstracts and science books. We have all been partners in the agonies of achievement; we have all been co-pilots on the journey.

---

* Publications on similar subjects are often set in place very much like the next stage of a construction. If you are building it properly, each level adds to the credibility. *The Journal of Environmental Quality* was certainly a major step in that direction.

Even though I had managed some excellent project partnerships at Madison and Green Bay, the preponderance of my significant research began at CRREL, starting with those first two years in 1975-76. And it helped to propel me into some of the most fascinating and productive discoveries of my life, including some forays and research conventions in the next couple of years that took me to Canada, Israel and the Soviet Union.

In fact, from the time I arrived at CRREL and Hanover, New Hampshire in the late winter of 1975, my life seemed to kick over onto a fast track with a series of adventures, one after the other. For a while it seemed as if I were leading four lives at once—as a scientist, an entrepreneur, a researcher on a magical mystery tour and a bachelor doctor.

Still, as we know, the only thing constant is change. And my status in one of those areas was about to be altered for good.

CHAPTER 26

THE BACHELOR DOCTOR

Part of my challenge as a workaholic has always been the willingness to find any time for myself. Even when my ex-wife Marcelle and I were granted a divorce in Wisconsin, I was still working to hold onto all my identities—doctor/dishwasher/soil scientist and father—and still have a moment or two to myself.

As anyone who's been through a protracted divorce will tell you, the wounds don't heal quickly. Especially when you have children, you still have family ties that lock you in emotionally, and lock out any other spaces in your heart. So, I resorted to my one safe-haven—that of the dedicated soil scientist, convinced that my work was the only thing that mattered. Having said that, I've come to learn that "life" always finds away; that a wounded heart will heal. And a need for life's little pleasures brought a crack of light back into my days.

I believe it was the comedian/show-business mogul Lucille Ball who said: "If you really want to get something done, give it to a busy man." I tend to believe this is true, because somewhere along the line I even managed to make extra time for myself. Even when I was back in Madison in 1973, working two jobs and spending weekends commuting to Green Bay to see my children, I was eventually able to rearrange my time to function as a normal human being, and a bachelor to boot. And yet it wasn't an easy thing to return to that "singular state of mind."

As is the case with most young men my age, my social urges eventually prevailed, and I started dating again: younger women, I think because I wasn't looking for anything that might present a commitment.

Living in Madison was an interesting experience. Since it was not only the base-home of the University of Wisconsin but also the State Capital, there was always the opportunity to meet an interesting cross-section of people. One of the women I met and dated for a while was a Jewess from Eastern Europe (Belarus, as I recall). She was very young, beautiful and headstrong. And though the relationship became serious for a short time, it was the kind of pairing that was doomed from the start. This was due in no small part to both ethnic and religious differences.

She still had family ties to Israel, and Israelis somewhat justifiably still had a siege mentality where anyone from the "Arab" world was concerned. At the time, Egypt still symbolized all nations opposed to Israel's right to exist. So she never could have taken me home to "meet the parents," as it were. The other issue was, of course religious. Orthodox Jews require conversions if you're to marry into the clan. As a devout Coptic Christian, I wasn't about to abandon my roots, and of course neither was she. And so, with so many points of disparity—race, religion, xenophobia and age—we both realized that we had no place to take our relationship beyond the pleasure of one another's company.

A short time later, I started dating a lovely lass of Irish origins—a student filled with youthful enthusiasm, ginger hair and a genuine gift for living. She was twenty and I was thirty-five. In a way she was still a child, and I had lived a couple of lifetimes already.

It is funny when you reach a certain age that one thing becomes abundantly clear: Drama is the province of the young. If you're dating someone who is just a few months out of teenage, you're bound find their world full of problems that you resolved long ago. Especially in the realm of romance, myopia reins supreme, and try though you may you cannot convince them that certain emotions will pass. So, since I had experienced enough emotional Ping-Pong in my own years of marriage, I decided to cut this relationship loose as well.

These "dating" experiences were something I had written off as just a part of becoming my own man again. But they did help give me perspective where my own former marriage was concerned. I was relieved to find that, as an ex-wife, Marcelle had been very accommodating when it came to my visiting privileges. And nothing brings a man greater peace of mind than access to the children that he adores.

That was the reason it made it all the more difficult for me to leave Wisconsin and sail uncharted waters to some mysterious cluster of communities 1200 miles to the East.

The matter was complicated further by the fact that some friends and relations from Marcelle's side of the family tried to get us back together, if for no other reason than for the sake of the children.

Even though we never even broached the subject once I moved to New Hampshire, somehow along the way a kind of healing took place between us. We shared a son and daughter in common, and that would always tap into the goodness we found in one another.

By the time I had reached Hanover, New Hampshire and my job with CRREL in 1975, I came to this town convinced that I might remain a bachelor indefinitely. I had no need to remarry any time soon, and I was too busy to date. So, by the time I had been there about six months, I was all work, and had little use for the frivolities of life.

I had a few dates from time to time, but the locals lacked the sophistication of those varied and interesting souls in Madison. They were nice enough, but this was no place for a single man.

Hanover was a college town of about 10,000 people, over 50% of it owned by Dartmouth College and more or less dependent on the businesses that served it. CRREL was the only other major enterprise in that vicinity…along with a noted Medical center affiliated with the University. Hanover and nearby Lebanon were predominantly Caucasian with a WASP white-bread sensibility that barely tolerated dark-skinned aliens such as I.

I got on well enough with the locals, and casually dated a handful of women from time to time. But I was simply to career driven to ever take the place seriously. Besides I was now travelling a great deal. And this town, in many ways, was starting to become little more than a place to hang my hat—a stopping off place between home and my lab work; that an little else.

That was good enough for me, I thought. I had plenty to do. I might not find a bride again, and that would be OK too. After all, I didn't know that many happily married people. It started to seem to me to be the ultimate social paradox; necessary to extend the bloodline, that and little else.

Then one sunny day in August 1975, I realized I needed a haircut. And someone recommended a friend…

CHAPTER 27

THE SECOND TIME AROUND

At that point in my life I have to admit that my hair wasn't the easiest to cut. Even more than now, it was an era of long hair, sideburns, muttonchops, mustaches and beards. Although my hair wasn't long, it was coarse, thick, curly and more than a bit unruly.

After accepting a referral from a co-worker for a hairdresser I found myself entering Mr. Brad's and meeting my future wife, Bonnie. Although I wanted to ask her out on a date right away I hesitated out of respect for her workplace. What's more, I was so caught up in my thoughts about her that I completely forgot to leave a tip!

All through the next week Bonnie lingered in my thoughts and I wrestled with the decision to call and ask her out. Finally, I got up the courage to call her at Mr. Brad's and ask her for a date. We had our first date at a Lebanese restaurant called Lander's. As we ate and talked easily with each other I lit a cigarette, but after seeing that she didn't smoke I put the cigarette out; I was that intent upon making the right impression.

When Bonnie Palmerston and I met she was 33, naturally beautiful, and had two sons from her first husband who had recently passed away suddenly from a brain aneurysm. The boys, Tony and Bruce were nine and five years old respectively. At the time it was still fairly uncommon to have blended families but after dating for a while we found that everything had just fallen into place so perfectly that we were meant for each other.

One day while we were sitting in front of Hanover High School and her sons Tony and Bruce were playing on the grounds in front of us, I could sense that Bonnie was trying to tell me something. Agonizing, after a bit of a pause, she finally came out with the fact that she had also been seeing another fellow named Armand even before she had met me. I was really quite disappointed to see that she was trying to juggle some mixed emotions about whom she would continue to be with.

To my surprise, when Bonnie announced that she might have to stop seeing me, I began to weep. It was a spontaneous outburst, almost reflexive on my part. But I'll never forget the moment—because it was then that I realized that I actually loved this woman—not only loved her but also might want to spend the rest of my life with her. I could tell that she too was touched by the honesty of my emotions; that the tears flowed despite my attempts to hold them back.

I had no idea before now that Bonnie was really torn between the two of us. I could tell that she had been touched by the sincerity of my feelings. And that had made the decision all the more difficult. This was not a choice she could make overnight, she told me. And so she went home to try and figure things out.

What I didn't know then but found out later was that Bonnie had decided to discuss her situation with her two closest friends, Linda and Janet. Linda Facto in particular had been BFF *(Best Friends Forever)* with Bonnie since they were 5 years old.

"Alex is the one for you," Linda told her. "He's kind, smart and easy to get along with. He obviously cares a great deal for you…and he's winner! He's got a doctorate, a great career, and he's solid. He's the kind of man who will always be there for you."

When Bonnie got back in touch with me to inform me of her heartfelt decision, I couldn't have been more pleased. And we decided to merge our lives and our families shortly after that time.

Bonnie was living in an apartment in a small farmhouse in Lebanon at the time. I was still renting my one-bedroom house close to my work in Hanover. Bonnie asked me if I was interested in moving in with her to "share the expenses."

I wanted very much to be with her, but still had to think things over. Despite my deep feelings for her I felt very much like Caesar crossing the Rubicon. Once I had stepped into that river, there could be no going back. I was committing to something long-term. There was simply no glossing over the issue; this was IT! I thought about it long and hard, and finally jumped in; I moved in with Bonnie in the spring of 1976. It was the right decision.

Over the next few months, our relationship grew until it felt to both of us as if we had become inseparable. Originally, I had wanted to find a woman who was down-to-earth, attractive and with values. And yet God had seen inside me and had found a deeper need. So He brought me someone who was so much more than I could have asked for…or ever hoped I could find. Soon after Bonnie and I came to live together, I also began to realize that I had merged my life with a woman who was kind, loving, compassionate, generous and extremely loyal to all those she loved. Bonnie always put her loved ones first, which made her even more beautiful in my eyes.

Our relationship grew over the coming months to one of both deep affection and trust. So it was during a 20-day business trip to Russia that I decided to propose marriage. Sometimes the day-to-day lives we share blur the core of our feelings. But the distance and the pause to reflect on my life gave me the time I needed. I made a telephone call from Moscow and asked her if she would marry me. To my delight, she answered with a very passionate, "Yes!"

On the way back, I purchased a diamond from the Moscow Duty Free Shop. I changed my return ticket and flew over to Egypt for a few days to ask my parents for their blessing before I came back to the US.

When I got together with my mother and father and with my sister Fayza, I think they could see the change in me—the light in my eyes, the lifted vibration, that joy that falls like grace on a man when he has found that other part of himself. If words alone would not convince them, I'm sure my energy did. It had shifted, all to the light; it had become tangible.

My brothers wished me well, as well. I think they'd remembered my first decision, and some sarcasm might have crept in that "Anything would be an improvement." I assured them that this time I was doing the right thing.

As I was leaving my family in Cairo it had suddenly struck me that the woman I was about to marry had a great deal in common with the women in my own family, my cherished mother and sister. Love and loyalty and a generous heart—these are the things that matter. Bonnie had all these things in spades.

There is a love ballad called "The Second Time Around" made popular, I believe, by Frank Sinatra. And the words go something like this: "Love is lovelier the second time around … Much more beautiful with both feet on the ground." Well, this was the second time around for me, and yet for the very first time it was my own choice, with both feet on the ground.

Bonnie and I decided to have a small wedding in our Kinne Street home in Lebanon on October 15, 1976. We didn't have much money saved after purchasing the house. Tony and Bruce were living with us; so we wanted to keep it simple. Until now I still wasn't familiar with Bonnie's family. Bonnie has always remained close with her sister, Mary. Otherwise, I came to learn that she had been adopted by the family that raised her, but she had become somewhat estranged from some of her "step" siblings as she grew into adulthood. It was clear to me on the day of our marriage that this wedge driven between them before I had even met her would continue to grow.

On the day we wed only two members of Bonnie's foster family came to our ceremony. Her adopted mother declined our invitation, choosing instead to go and play bingo. Although I didn't feel offended by these people I knew that Bonnie was crushed, and I became determined to protect her from that type of emotional pain. We truly became our own family, depending on each other.

Bonnie looked beautiful on our wedding day, wearing a dress that I had bought for her on one of my business trips for CRREL. She did her own hair and makeup and spent the whole day with me and her sons, cleaning the house and preparing food for our guests. We bought a small cake and had a Justice of the Peace perform the ceremony.

At this point, I have to confess that although I had strong hope that Bonnie and I would spend the rest of our lives together, I was still hesitant to jump in with both feet financially speaking. Given the terms of my divorce, I was still paying alimony and child support. Having just bought my first property and now about to remarry, I was experiencing the pride of ownership and a touch of paranoia. So, I felt as if I needed to protect that part of my life.*

This decision had probably been enhanced by a couple of things: First New Hampshire was a community property state. So, upon divorce, the wife got half of whatever the husband had. Second, there had just come about the famous Lee Marvin/Michelle Triola Marvin "palimony" suit, wherein she was awarded half of everything he owned. This singular event hastened something called a "prenuptial agreement" to the forefront of public consciousness. Everyone was doing it, and that legal procedure had not been lost on me.

Early on in our engagement, I asked Bonnie if she was willing to sign such an agreement. If a divorce occurred, each party would be entitled to keep the stuff they owned before marriage. All the furniture belonged to Bonnie and the house was in my name alone.

Bonnie readily agreed to the document. And what I could see unfold was a beautiful human being who held no attachment to "things." (In that way, she was ahead of me; perhaps ahead of us all.) In later years, I realized that this was the one woman I could trust with my life. A few years into my marriage, I tore the agreement up and threw it away. That would begin yet another chapter in our personal and business partnership.

---

* At that point in my life I was 38 and still covered up in a blanket of financial obligations. I was not only paying alimony to Marcelle and child support for Niveen and George, I was also sending part of my monthly paycheck and fees back to help my family in Egypt. Now I was facing the addition of a whole new family—a wife and two more children. So, just based on the math alone, the prospects were daunting.

CHAPTER 28

A MATTER OF INTELLIGENCE

The years between 1975 and 1977 were significant for me for all the right reasons. The research papers my colleagues and I had pulled together on wastewater management and renovation had generated both publication and peer review among environmental science groups on a worldwide basis.

Understand that the universe most scientists dwell in is very small indeed. And though the environmental implications of issues we often address are vast, the specifics into which we delve usually involve a relatively narrow scope of interest. Or do they?

What I was soon about to find out was that everything we did in the world of CRREL and the universe of ACE was going to be scrutinized by the CIA, the FBI and the National Security Agency (NSA)…and looked upon as possible issues of national security where the USA was concerned. That is hitting you with a lot of acronyms, I know. But I have since learned that acronyms are just shorthand that makes it easy to remember much longer sounding names that are occasionally a little bit scary.

During the period from about 1975-'78 the world was involved in all kinds of cold wars and unofficial conflicts scattered all over the globe. The Vietnam War was still fresh in everyone's mind. Southeast Asia was coming undone and remained in constant turmoil. Cambodia had been overrun by the Khmer Rouge. The Sandinista (Communist) Rebels were churning up a Civil War in Nicaragua. And Egypt, in 1973, had launched the Yom Kippur War, counterattacking Israel in ways that had enabled them to regain a large portion of the Sinai Peninsula that it had lost five years earlier.

Initiated by President Anwar Sadat, The Yom Kippur War lasted nearly a month, involved a coordinated military alliance between Egypt, Syria, with tactical support from four other "Arab" nations, and came just close enough to successfully repelling the Israelis to cause the US to ramp up its military and the USSR to rattle its sabers, bringing everyone to the brink of yet another global conflict.

US Secretary of State Henry Kissinger had softened the blow by eventually initiating Resolution 339, thus setting off a chain of "Peace Conferences" that ended at the Camp David Accords in 1979. During that five-year period, relations with the US and the Middle East worsened, a chain of oil embargos caused a spike in gas prices in the US to a whopping $1.29 a gallon, long lines for gasoline and the fall of the Pahlavi Dynasty in Iran.

Around 1976 most of this was lost on me, to be sure, because I was now too far removed from the geography of conflict to have any of this nasty business affect me directly; or so I thought.

I suppose I probably should have known better. I was after all joining CRREL, a chain-link sibling to the United States Army Corps of Engineers and a research branch specifically set up as a cold war technology to create permafrost alternatives and cold weather tactics should the US Military ever need them. And even though most of our research was being applied to civilian criteria for "peaceful" purposes, we were still in a military mentality, evidenced by the fact that the head of CRREL when I first arrived there was an ACE "Bird" Colonel.

Nothing underscored this more emphatically than a point I reached about ten months into my new job, and it involved a statistical abstract on our own wastewater management techniques at CRREL that I had submitted to the Israelis.

I had long been an admirer of the Israeli's command of their ecology. Their farming, cultivation and irrigation techniques were light years ahead of any other nation in the Middle East…and probably in the world. They had taken one of the most geographically challenged pieces of arid desert in modern history with less than 24% innate arable land and turned it into a model economy with more then 86% productivity from its farm potential—a veritable Garden of Eden of arbors, groves and cash crops. Much of this had come from their ability to utilize every bit of land, water and energy from available resources.

Since water was at a premium, the issue of Wastewater Management and Application (my specialty at the time) was of primary importance. So, we were very much on the same page.

I was looking forward to traveling to Israel for the exchange, and for a convention they had scheduled for me to attend. What hadn't occurred to me at the time however was that, since we were in a government agency with sensitive military information constantly at our disposal, we were required to receive official clearance before we could travel anywhere overseas.

Still, it came as something of a surprise when a CIA agent suddenly dropped in at the "underground" telephone room that Dan Leggett and I were using for an office. He was polite at first, but it didn't take long before he was probing into the fact that I (an Egyptian) had just presented a position paper for the Israelis and was now planning a trip there. After grilling me on the paper and my intentions for the content, he seemed to be satisfied that I was a Coptic Christian who had actually come to America partly because of the religious freedom here, and the fact that I cherished my US Citizenship and all the privileges it implied.

"Have you ever worked for another country?" the Agent inquired.

My response was one of pure reflex. "No. I'm from Egypt. How would I work for another country? And why would I?"

"Would you like to work for us?" came the query. "For Military Intelligence."

"In the first place, I've never served in the military," I answered. "I was rejected for the Egyptian military because I was considered redundant. And I was rejected by the draft board once I became a citizen here, because I was technically too old. I'm not really very good at military things. I'm not even good at sports." I don't even know why I had said that, but I was willing to do anything to dissuade him. It didn't.

"We have something a little different in mind," the CIA agent responded, implying that I might be useful as a possible operative for gathering intelligence in both Israel and for any upcoming trips scheduled to the Soviet Union.

We ended with him asking me to consider his proposal. At the time, I didn't know what else to tell him. All I could think of was that it would be the worst sidetrack of my life. I wasn't sophisticated in the ways of espionage, but I knew enough to know that, once you got on board this intelligence train, there would be no way to get off.

That night I remember talking it over with Bonnie, and expressing the fact that I didn't want to have this kind of dangerous double life. (I couldn't imagine anyone wanting it, in fact.)

I don't think I slept for more than a couple of hours that night, and the next day I went to our CEO at CRREL, The ACE Commandant and Director, for some clarity on what had just happened to me the day before, making sure to get across my utter opposition to such a proposal. Somehow it didn't surprise me that he knew, but it relieved me greatly that he gave the answer he did.

"I know. I told them to leave you alone. That you were not the right man for that kind of thing," he answered. "I also told them you were too valuable in the work you were doing for us to be disturbed with all this other stuff."

The Colonel was good to his word, and from that point on I never heard from the CIA again; at least on that issue or for any other "intelligence" agenda. It was however not to be the last incident my Egyptian roots and my affiliations to the Middle East created for me.

Once on my way back from an Environmental Science convention in Canada in 1978, I ran into another occasion where my unwitting lack of discretion caused me some inconvenience. Ironically, since the implementation of the North American Free Trade Agreement (NAFTA) Americans have had to present passports to get into and out of Canada. Back in the 1970s, however, they did not. And yet, upon leaving Canadian customs, I was asked about my destination, and I told them I was going to "Lebanon."

Well... the Canadian Customs official took one look at me, at my "Middle Eastern" appearance, plus the name on my ID, subsequently assumed that I was going to Beirut or something and immediately demanded to see my passport. Fortunately, I never left home without one and just happened to produce mine (replete with my US Citizenship), after which point I was finally able to explain that the "Lebanon" I was referring to was actually a borough in New Hampshire about 9 miles from Hanover and Dartmouth College.

The irony in all of this was that, like so many places in the world, Lebanon was in the midst of a Civil War between Muslims and Christian Phalangists—something that the Canadian officials seemed concerned with that I was not.*

Not that I was indifferent to the strife in the Middle East. It was just that I had other things to do with my life.

Though our collective accomplishments in Wastewater Management, Applications and practical uses from 1975 thru 1977, I was now being invited to present papers and speak at conferences all over—from Canada to the Soviet Union. Without going into every single paper and the scientific details inside them, our work was all being directed toward three major objectives: 1) converting vast quantities of toxic wastewater into good citizens first by fingerprinting them and second by transforming the heavy metals and trace elements inside them through proper "natural" balancing; 2) applying that balanced wastewater as a positive growth medium through three natural filtrations, thereby restoring the ecology, creating healthy new farming techniques and end the collision course with an inevitable global drought, which is still a danger; 3) the reuse of wastewater not only as a positive nutritious growth medium but also as a versatile, potable form of drinking water (which currently has the ability to exceed most bottled waters in terms of purity and natural mineral content).

In 1976, a paper I had officially co-authored, netted an award from the US Military Academy at West Point, and got me an invitation from noted Professor

---

* The Christian Phalange in Lebanon was originally funded and armed by Israel, who had the mistaken notion that they would ally to Israeli interests in the Middle East. To their surprise the Phalangists were fanatically right wing and even retained a Nazi salute until the late 1990s. (Further evidence that one should always be sure to vet one's associations.)

Albert Page to speak at an International Conference on Trace Elements in the spring of 1976. That in turn, albeit indirectly, led me to a trip to the USSR to represent CRREL for what had become a fairly famous Science Exchange Conference in Moscow.**

I have mentioned this before but emphasize it again: In the world of science, the realm of ideas, there are no borders, barriers or nations. We are fellow travelers in the vast universe of discovery; ours is a brotherhood of shared information. (Besides, at this particular convention, I had other things on my mind: one of them was proposing to my future wife. The other was arranging for a side-trip to my homeland.) Unfortunately, governments do not share the collegial sensibilities that most scientists enjoy.

Through my longtime friendship with Egyptian friend and microbiologist, Hassan Moawad, I had arranged to meet with some colleagues in the field of soil studies and biogeochemistry while I was in Moscow. Hassan had studied there in the early 1960s, acting on an Egyptian/Russian exchange program as a lecturer and research assistant. He had cultivated a number of valuable associations while he was there and had made certain that I would be able to share in those contacts and the exchanges of information they might provide.

So I suppose I thought it a bit bizarre because the whole time I was in USSR, staying at the Russia Hotel near Red Square, our every little move was being monitored by the KGB. Every day when I would leave, I was required to give my key to a "Floor Monitor" at the hotel. Later at night I would return to my hotel room only to find that it had quite blatantly been searched, that

---

** Murrmann, R.P. and I.K. Iskandar. 1977. "Land treatment of waste-water: Case studies of existing disposal systems at Quincy, WA and Manteca, CA." Land as a Waste Management Alternative, R. Loehr, ed. Ann Arbor Science. Ann Arbor, MI, pp. 467-488.

Baillod, C.R., R.G. Waters, I.K. Iskandar, and A. Uiga. 1977. "Preliminary evaluation of 88 years rapid infiltration of raw municipal sewage at Calumet, Michigan." Land as a Waste Management Alternative, R. Loehr, ed. Ann Arbor Science. Ann Arbor, MI, pp. 489-510.

Uiga, A., I.K. Iskandar, and H.L. McKim. 1977. "Wastewater reuse at Livermore, California." Land as a Waste Management Alternative, R. Loehr, ed. Ann Arbor Science, pp. 511-53 1.

my bed had been turned over, and that all the film from my camera had been removed.

I had heard it would be this way. (Even Hassan had warned me.) I just never thought that the journey would be so laced with drama. Then somewhere along the line it struck me: What if I had in fact taken the CIA up on its offer? What if I had decided to break from my destined path and had plunged instead into that hidden world?

How much different would that trip have been? What kind of person would I have become? Of course it was never a danger really. I am a man of science. Mine is an endless search for truth. And this was the realm of deception—of point and counterpoint—as alien to my field of endeavor as any I could ever imagine.

Years of research in laboratories have taught me that one should never deal in absolutes. This, I think, applies to life in general. The world would change in the next 30 years. We would lose our innocence as a nation. And I would actually volunteer to serve my adopted country in any way I could.

By and large the trip to the Soviet capital was quite professional and mutually beneficial to all those who attended. Nevertheless, I soon came to realize that, given the cold war dynamic still in place, and given my strategic position with a US Government Agency, I was going to be debriefed every time I went to the Soviet Union.

This was the world we had come to live in; such was the world we had created. I wondered if it would ever change. Then, when the Berlin Wall finally came down in 1989, I delighted in the fact, believing that somehow peace would prevail and the global community would finally come to its senses. Of course, I couldn't have been more wrong. Before long, it seemed to go even more insane. And soon enough in the scheme of things my patriotism would be challenged again.

It's hard to realize that it has already been 15 years, and yet there was a collective period of madness that struck America before, on and after September 11, 2001 that affected us all. And even though, at some deep spiritual level, I acknowledge that there are no accidents, I still marvel at the fact that strange and terrible moment in history happened on the very the day I had chosen to rent a car and drive my sister-in-law to Logan Airport in Boston.

So, there we were in an "Arab" Panic, trying to make sense out of the madness and the fact that there had been a coordinated suicide pact to destroy as much of our Nation's substructure as was humanly possible. (23 terrorists, including 14 Saudis, had just taken down the Twin Towers in New York, had just crashed into the Pentagon and killed 189 people there, and had just seen United Airlines Flight 93 shot down apparently on its way to crash into the White House.) And some of this had launched from Boston; so it was in fact a point of concern.

We were a nation under attack; and we were also overreacting. The FBI, with egg on its face from ignoring CIA warnings of a possible attack, was in a mad scramble to seize anything from any Middle Easterners that might even look suspicious.

That spilled over onto me in an almost farcical way when they showed up at our house a few days after 9/11 to check on my whereabouts that day. Apparently, they had seized the computer hard disks at all local car rentals up to an on September 11, 2001 and my (Egyptian Arabic) name had been flagged.

It's as sad to note as it is amusing that our intelligence agencies are often utterly inept when it comes to knowledge of the rest of the world. All the mystique of 007 James Bond or even (Tom Clancy hero) Jack Ryan are pretty much the stuff of myth and fiction. It's a plodding business. And most agents tend to dress and act like insurance salesmen. Still, these men who called on us had badges and grave concerns, especially about why I would be going to a launch point like Logan Airport on 9/11.

So I had to sit them down and recite my standard socio-ethnic explanation along with my personal resume that, I was Egyptian, I was not a Muslim but a Coptic Christian, and since we were being subjected to religious and political intolerance ourselves we were the last people in the Middle East inclined to terrorism.

What's more, I also had to enlighten them to the fact that many of the names on their long list of suspects were all my relatives named "Iskandar," including my brothers, my nephews and their families. I also underscored that I had worked for CRREL and the Army Corps of Engineers for over 25 years. And they could check my high level of government security clearance (which they should have done in the first place).

It certainly helped my cause that the local Chief of Police in Hanover and Lebanon also gave me a vote of confidence, implying that I was something of a local pillar of the community.

"Alex is a good guy," the Chief of Police told the investigators, an endorsement that put them at ease. And after a brief time they went away.

This entire chain of events made me feel somehow violated, but not in ways that anyone might imagine. What bothered me most (saddened me really) was not that I was being sought after because of my Egyptian name or my roots in the Middle East. What troubled me more than anything was the realization that our Intelligence Agencies were so poorly informed and ignorant of what the true roots of terrorism in this country might actually be. So, living up to my credo that Crisis Creates Opportunity, I decided to do something about it.

Working through local law enforcement in Hanover and Lebanon, I got in touch with the head of the regional FBI headquarters in Manchester, New Hampshire to offer my *pro bono* services as a translator of anything written in Arabic, a cultural screener, and a resource to check out any fishy looking websites written in Arabic with possible references to any Al Qaeda or Radical Jihad sleeper cells that might be cropping up.

If this all sounds a bit paranoid and cloak-and-dagger, I might use this moment to suggest a reality check. As of 2002, there were estimated to be about 90 sleeper cells and at least 12 traceable Jihadist camps in the United States, many of them training actual combatants in sabotage and military tactics; most of them on the East Coast. (Those numbers have easily doubled by now and, despite stepped up government monitoring, they have not gone away.)

So part of my volunteer intelligence services to the FBI comes as a US Citizen and property owner in Lebanon, New Hampshire. It may sound pretentious. (I don't mean it to be.) But this is my country too, and it seems to me that we all have responsibilities to the communities we live in to do our part in any way possible. My command of Arabic and my awareness of the cultural Rubix that goes on in the Middle East have both enabled me to make some small contribution. (I only stopped doing this recently, but would again if asked. It's just that there hasn't been a call for this; some cause for gratitude.)

Winston Churchill once ironically noted: "Democracy is the worst form of government except for all other that have ever been created." He also noted that it is only as safe as vigilant citizens make it. I agree on both points. Citizenship is occasionally hard work. And yet this country we truly love is worth it.

CHAPTER 29

PUBLICATIONS: A BRAIN TRUST

I've already been warned in advance about this section of the book—not to make it some pedantic treatise on scientific data, facts and figures that few people would understand and even less would care about. To my surprise, most of these entreaties have come from my colleagues in the vast community of environmental sciences, including some of my closest friends.

In a way, I shouldn't be surprised. There's no real false modesty involved. They just realize that the language of science can be difficult, driven to detail and generally boring. But sometimes the results can be exciting and the ways they ultimately serve this planet occasionally even earthshaking. So let me dwell on those paradigms for a moment.

Meanwhile I can use this time and space to honor some of these extraordinary people who have shared some of these journeys of discovery and—I hope—make some of what we've done be interesting to you.

I was originally tempted to label this chapter "Soil Science for Dummies." And though it is still something of a temptation, it would be both demeaning and inaccurate to do so. I will however offer this brief primer in what we do to show you how one step of scientific research and development inevitably leads to another until the ladder is complete.

It's a short ladder in this book even though it covers about 40 years worth of discoveries. But for any of you who have seen the whole history of the world in sixty seconds on a video, this may be the print equivalent of that.

I know I have covered some aspects of Soil Science prior to this. But perhaps I should add a bit more perspective about the role it plays and the scientific and ecological umbrellas that now cover it. Although it may have not specifically started out that way, Soil Science is now considered a branch of Environmental Science.

Environmental Science is what is called a "Multidisciplinary Academic Science" in that it covers everything from ocean pollution to overpopulation, from global warming to zoological depletions.* Technically, it covers geology, marine biology, zoology and atmospheric science. But frankly, by now its umbrella fans out so broadly over everything that in reality it covers nothing. So saying one is an "environmental scientist" is very much like saying, "I am a Christian," or "I am a musician." One must then invariably ask: "Great! What instrument do you play?"

Soil Science might easily be looked upon as the rhythm section in the concert with Nature. (Consider water sciences—marine biology, aquaculture, oceanography, and hydrochemistry—to be the strings and air sciences to be the horn sections.) It all comes down to earth, air and water. And, they are certainly interactive. Everything begins there.

Soil Science literally focuses on every aspect of the soil. And studies of soil science in a way become a morality play. You may start out with mere intellectual curiosity about the soul and content of the earth and the soil that inhabits it. (And some academics just choose to stay there.) But eventually you end up choosing sides about where your knowledge will take you.

You may treat the earth and the soil that comprises it as assets to be exploited—to be optimized to serve us. (That is the role of most Agra and

---

* Environmental science is now one of the most popular courses of study at every major university across the US. But unless you are going to specialize and truly take on the science mantle, the Environmental Science label merely becomes fashionable—something akin to a Liberal Arts major, providing no particular set of skills once you graduate.

Chema giants…and most governments.) Or you may approach this relationship we have with Mother Earth as one of resource and restoration, acknowledging it as a finite source that we have a responsibility to understand, to facilitate and ultimately to restore. In the world of pure metaphor, that is the high road, and the one I hoped that my associates and I have been instrumental in taking.

I know not what course others may take. I can only speak for myself. But the more I have involved myself in scientific research and development over the years, the more strongly I have become driven to result—specifically the "development" side of things. For me, that means applying our scientific knowledge in the art of conversion.

Soil science is really quite practical in that by understanding the soil, defining its content and applying mathematical models to it, we can put it back into healthy balance. Human traffic, chemical influences of commerce, over-farming with the pernicious cycle of pesticides and chemical fertilizers, plus the destructive traffic of both fossil fuels and the machinery of war have so distorted the soil we rely on that—left on its own without correction and repair—the earth as we know it would eventually end up poisoning all other life on this planet as surely as a bullet to the brain.

If that sounds a bit melodramatic, at least understand that it would do so in a dozen different ways, from contaminating our food chain to killing our wildlife (especially aquatic wildlife and fish) to intoxicating our water supply.

So much of this is about understanding, defining and applying the mineral balance in the earth's terrain, especially the soil from which grows all life. Minerals—especially trace minerals—are essential to all life, provided they are in balance. This is especially true of nitrogen and carbon balance (but also applies to all minerals). *Carbon* is the basis of all ground life, but too much carbon gas in the form of carbon dioxide can be poisonous and life destroying. *Nitrogen* is considered the most essential element for healthy plant growth and nutrification, but only if kept within appropriate ranges of pH. Minerals such as *calcium, copper, iron, zinc cadmium* and *boron* all have important roles to play. *Mercury* and *lead* may be present but in highly restricted amounts. And the presence of such elements as *arsenic* and *mercury dioxide* can prove to be so dangerous at certain levels as to be classified as "toxic waste" that will require either isolation or removal.

This is oversimplification of course. There is much more to soil science than this. Still, the same rules apply. And the constant need to "balance" the soil and keep it healthy is a struggle in chemical strategy just slightly short of becoming a battle plan.

In the world of academic science, we were very often funded to isolate and define the challenges the air and water brought to the earth around us. When I joined CRREL, the parameters were reset considerably because both construction, massive land movement and the fuel and ordinance that military operations such as ACE brought to the table often created a kind of "toxic soil shock" that we were literally hard-pressed to study, define, remedy and solve, often under the pressure of shortened timetables.

This frequently required funding, sometimes on a massive scale. And so, rather than see this as a burden or feel the pressure we were often under, I applied my Iron Law of Personal Empowerment: *Crisis Creates Opportunity.*

So before too long, I became known as the guy who would plunge into any challenge set before CRREL and present both the brackets of alternatives and the budgets we would require to accomplish our objectives. In other words, I more or less acquired a reputation for becoming "the Money Guy," inside CRREL. (After all it was the Government. Not just any government: the US Government, the one that didn't mind throwing money at a challenge if there was truly the chance to solve it. So working up cost analyses only seemed a logical part of the process.)

As I had mentioned earlier, our first several projects at CRREL were devoted to Wastewater Management renovation and control. I reiterate what was at stake: that was ultimately conservation of the medium itself. Because the way it was set up in most municipal and rural runoffs was a matter of maximizing water usage without wasting it.

Having made numerous studies of the three major ways to manage wastewater—through overland flow, slow rate and rapid infiltration—and having broken down the reagent miscues and remedied the balance of the minerals, we still had to unlock more of the keys to effective wastewater management.

What we had to focus on was effectively advancing wastewater as it moves through the soil as a natural alternative to spilling it off as pure waste. To meet these criteria, we recreated soils in greenhouse conditions modeled after those germane to the native earth. Then we would extract samples at different depths in the soil.

To achieve this we went no farther than taking samples from wastewater from the town of Hanover's treatment plant, added balanced chemicals and eventually came up with the proper formula, setting up formulations that actually turned wastewater and sludge from treatment plants into effective healthy growth media for Agriculture—mathematical conversion models we were able to apply in different test areas from New Hampshire to Australia.

Understand at this point that this is a gargantuan challenge that exists in every major community in the world—from the sophisticated farmlands of Middle America to the desertification challenges of the Middle East to every developing country from Namibia to Nicaragua. So unlocking all the secrets to this took more than one aspect of Soil Science. It often takes the combined effort of agronomists, microbiologists, geomorphologists, engineers, physicists and mathematicians (just to name a few of the disciplines involved).

Solving soil science often involves a brain trust. And if I have acquired another talent over the years, besides the building of mathematical models, it is my ability to recruit the "best and brightest" professionals to partner with on the pathway to completion.

I have always been especially blessed in my career to work with some of the finest minds in chemistry, environmental science and soil sciences. Up to 1977, I was fortunate enough to blend my professional energies with the genius of Keith Syers, the prestigious aegis of field leaders such as Marion L. Jackson and friends and associates such as Dan Leggett. What I was about to put together was a dream team of some of the finest physicists and mathematicians in the field of Soil Science, including Dr. Marvin Schaeffer and especially a countryman of mine, Dr. Magdi Selim, who at the time was joining the Department of Soil Sciences at Louisiana State University (LSU).

By this time, we had examined so many ways to maximize the use of wastewater and eliminate the throw-away mentality that had wasted what had to be a small ocean worth of useful, potable water and potential growth medium for a virtual continent of crops, farms and forests. We had managed to optimize the effects of groundwater on minimum land and had done so without contaminating either medium.

What still remained to do was to reconcile the fact that so much wastewater was overloaded and still contained a quantum dross of heavy metals. Along the way to a final step in the resolution of the issue, I had discovered that nitrogen was the limiting factor. And organic nitrogen (the potential "good guy" that, with proper balance, holds the potential to nitrify the native earth) was, in fact, so overloaded that it was becoming the causal agent that was contaminating the groundwater, simply because past a certain level "organic nitrogen" could not be absorbed and was in effect creating dangerous nitrates in the soil. Rather than look upon this as some abstract scientific notion, we recognized it as so much more and that it had, in some nations in Eastern Europe, become environmentally lethal, causing what was coming to be known as *Blue Baby Syndrome*.**

Seeking a formula to reset the "balance," I had built the theoretical model to establish my axiom as well as a strategic plan to follow it, but I needed the ultimate mathematical matrix—the specific figures—that could to validate them and practicalize the result. For that, I recruited Dr. Schaeffer and Dr. Selim to help me solidify my findings. My new working associates proved to be an excellent fit—especially Magdi Selim.

** Among other genetic sources, *Blue Baby Syndrome* is often attributed to high levels of nitrate contamination in groundwater, resulting in decreased oxygen-carrying capacity of hemoglobin in babies, ultimately leading to death. The groundwater, frequently in the form of wastewater too rapidly returned to the water cycle, can be contaminated by leaching of nitrates generated from fertilizer used in agricultural lands, waste dumps or pit latrines.

There are those professional partnerships that actually become transformative in that they generate a professional synergy you only experience two or three times during your career. I certainly made that kind of connection with Dr. Selim. We enjoyed not only the bond of a shared homeland and similar upbringings, but also a high degree of motivation to delve into a number of undiscovered topics in the area of Soil Science.

Beginning with our collaborations on wastewater and our nitrogen models, we also shared passions and pursuits in a number of areas over the next two decades, including studies and resolutions in heavy metals, special projects on military explosives, residual ground pollution, and *The Fate and Transport of Heavy Metals in the Vadose Zone*. (The latter is the title of book we co-edited and compiled, one of six we ended up creating together over the years.)

At the time I first met Dr. Selim in 1978, he was a member of the Soil Sciences Department at LSU (Louisiana State University) in Baton Rouge. And his work, including the mathematical construct he created from it, was so impressive that I tried to hire him away from that prestigious college on more than one occasion.

He politely declined each invitation. I certainly respected that. When a man has a true sense of his own destiny, he is able to resonate with one place and time in his life more than any other. Dr. Selim, by his own admission, just felt more comfortable in the atmosphere of university life and his need to teach "the promising minds" that came to him for knowledge, guidance and mentoring.

Even though we were both aware of the insidious degree of politics that often goes on in the world of academic science, Magdi remained convinced that this was where he was meant to be. He also acknowledged that he would be able to maintain that professional balance and sense of security that kept him operating within his personal comfort zone.

When I reflect upon it, it was a very wise decision on Dr. Selim's part for a couple of reasons: First, the world of science is almost always impressed by diversity of source and independent perspective. Had we both been working for CRREL and ACE, our partnership would have probably lacked the professional leverage we were able to achieve through this cross-pollination of disciplines. Second, I believe the geographic distance and different *in situ* study environments might have availed us of an even broader base of research and exposure to information than we might have experienced coming from "the same place."

As it was, Magdi Selim and I ended up meeting five or six times a year for nearly a decade, including about six weeks almost every summer, and to say the least, we certainly managed to get a lot done during those times...

Working together with a sense of urgency we might never have experienced otherwise, Dr. Selim and I were able to produce more than two dozen position papers and abstracts—including 10 in 1978 alone—a half dozen books and anthologies on everything from wastewater to heavy metal analysis and resolution, to some of the most challenging environmental predicaments in modern history.

Mind you, these are projects that involved seemingly everyone with energy, environmental and agricultural concerns, encompassing everything from the EXXON *Valdez* incident at Prince William Sound, Alaska to the fallout from the Nuclear Disaster at Fukishima, Japan—and projects and initiatives instituted from a virtual alphabet soup of critical government agencies all over the world, including The Department of Energy (DOE), the EPA, the USDA, OSHA, ACE and the UN.

As Mr. Spock might say, "It's only logical" that, with all that Magdi Selim and I shared in common, an uncommon friendship would arise out of so much professional synchronicity and respect; and it did. Marriages, children's weddings, birthdays and holidays—we ended up sharing so much of one another's lives, that we truly came to define the cliché: "brother from another mother." (We've even been there for one another's health challenges: my heart issues in 2014 and his lung transplant in 2011. By every other definition of "family," we have become all that. All this from two kids from the ghettos of Cairo—one a Coptic Christian, the other a Muslim—to this very special time when we get to continue to share the fruits of our labors, including a foundation I helped to found that Magdi has helped to perpetuate.)

One of the greatest by-products of my longstanding relationship with Dr. Magdi Selim has been our mutual effort on behalf of founding, funding and presiding over a non-profit entity called the *International Society of Trace Elements Biogeochemistry and Environment,* in 1994. More than just another convention where scientists go together to share ideas, ISTEB was set up as part of my personal passion for seeing to it that we would actually be able to bring the benefits of our scientific synergy to the countries that needed them most.

Set up as a 501(c)(3) non-profit, ISTEB not only hosts biennial meetings all over the world, including and a convention in Fuokoka, Japan (in July 2015), it also functions as a vehicle of scientific empowerment for developing nations such as Nigeria, Namibia, Bangladesh and India to help them optimize their limited "earth resources." In fact, we have set up an annual award to scientists and nations who are working to advance the understanding of Soil Sciences and Trace Elements in the development of their own nation's sensitive ecologies.

The premise behind our work is simple but driven to a point: All the scientific knowledge in the world is of no value unless it can be put to practical use in helping those nations and world populations who can benefit from our discoveries.

So far, ISTEB has made great progress in all these areas, but still has so much more that we'd like for it to accomplish. As past President and current Treasurer I still remain active in that pursuit. And we believe some of our best work remains to be done.

When I look back upon that critical span of years between 1978 and 1981, it seems almost incomprehensible that we were able to accomplish so much. Not only did we take on the seemingly insurmountable enigma of wastewater management and ultimate resolution, leading to the tome, *Modeling Wastewater Renovation* (still the "Wastewater Management Bible" to this day), we were also pressed to come up with correction models for groundwater pollution and remediation for the rest of the world. This was a "pollution solution" ultimately generated over a couple of decades, finally culminating in another set of models created for the rest of the world—especially in areas where a preponderance of industrial pollution had to be reconciled with the needs of immense populations such as those found in Asia, Africa and what we scientists refer to as Oceana (Indonesia, Australia and the Southeast Pacific).***

During the span of years, especially in the 1980s, we had also managed to venture over into energy and ordinance detection and resolution to "frozen ground" technologies that would dominate so much of my work in the coming years.

Add to that the fact that, for several years, I had been involved in developing alternative streams of income, including buying properties and learning about the demands of property management. Discovering with some delight all the ins-and-outs of real-estate, all the personal delights of a newlywed along with a professional reputation that I was starting to build, and I found myself having the time of my life. It had become such a constant rush that it felt like a mission in a Voyager capsule, fraught with new adventures every day. Frankly, no one could have enjoyed it more than I.

This was the way life should be, I thought. Nothing could stop me now. And yet something always does, it seems. That is life's lesson to us.

Then, just when everything seemed to be going on in a splendid kind of perfect order, I was thrown a curve ball that was about to give me perspective. I had just turned forty, and it was time for "a physical."

*** The effective evaluation, modeling and treatment to resolve these pollution challenges was the work of a dozen significant studies that I was able to compile into an effective summary, working with another soil science associate and biogeochemist, Dr. P.M. Huang (another University of Wisconsin association), summarized in *SOILS and GROUNDWATER POLLUTION and REMEDIATION,* published in 2000.

CHAPTER 30

SUGAR BLUES: THE DANCE OF DIABETIC DNA

The good news for me was the fact that, working for a US Government agency like CRREL, I was obliged by contract to undergo an annual physical. It was something I accepted as part of the job description and, in fact, welcomed the day or two devoted to it. First, the expenses of treatment and examination were entirely covered by the government. Second, given my family's health history one could never be too cautious.

In late October, shortly after my birthday in 1978, I went through the usual process of my physical only to find myself staring down into a terrifying set of medical data: My blood sugar level had bottomed out precipitously, indicating conclusively that I had tipped over into a level of *Type II Insulin Resistant Diabetes.*

Up to the time of my physical I had noticed an occasional drop in my energy levels from time to time, but since I had always been driven to work long hours, I sensed that the fatigue was mainly a side-effect of my workaholism—not all that far removed from a dread of poverty and a total adulthood of long hours devoted to my research and my side businesses.

At this point, I have to admit to having had a bit of a sweet tooth all my life, an addiction to good tasting food that I think all people who grow up hungry tend to find comfort in. And I had never spent a moment's time working out, exercising, playing sports or indulging in the new rage of the 1970s—fitness clubs and long distance running.

I had also been a 2-pack-a-day smoker up until 1975, and only a personal epiphany and the influence of my lovely lady had prompted me to drop that nasty habit.*

I was probably 15 or 20 pounds heavier than I should have been. But for a "driven" scientist and businessman, with erratic eating habits, who had just hit that dreaded milestone of forty, it just seemed part of the package.

Still, the diabetes diagnosis hit me like an uppercut to the chin for any number of reasons. And for a day or two I became deeply depressed because I looked upon it as a kind of death sentence; at least I did so originally.

Although I deliberately chose not to dwell on it, I had always been haunted by the memory of my aunts, Oginee and Galila and the personal hell they both went through later in their lives. They only lived into their early fifties. So watching them go crippled and blind, losing both their hearing and their ability to get around—dying by inches so to speak—seemed to me to be the worst kind of fate that could befall a human being. I knew that the side effects from diabetes had killed at least one of my siblings. And it had indirectly wreaked havoc on my sister Fayza in what had to be a Pandora's Box of illnesses she endured.

I knew diabetes tended to strike women three times as often as men. But up until the 1960s, there was still little differentiation between the types of diabetes—that there was a *Type 1* and *a Type 2*.

Proactive by nature, I have always believed that ignorance and fear are the coward's credentials. So I decided to arm myself with as much knowledge as

---

* I offer empirical evidence that Bonnie had a profoundly positive influence on me from the beginning of our relationship by pointing to something that happened on our very first date. Until that time in 1975, I had been a cigarette smoker—2 packs of Pall Mall Cigarettes for fifteen years at least. After our first dinner together, I pulled out a cigarette, lit it and offered Bonnie one. Without judgment, she merely declined. "I'm not a smoker," she said. Weighing the moment, I immediately doused my cigarette in the ashtray and never lit another. (Although I did smoke a pipe for a year or so.) So who is to say? Without even exerting an effort or trying to assert her opinion, Bonnie not only got me to quit smoking she may have even saved my life in the long run. It is something she's been good at doing, I have come to find.

I could about the subject. And I was also the beneficiary of some very good medical counseling.

As it turned out, the deadly dance of the diabetic had been more than ameliorated by the late 1970s. Medical expertise had exponentially advanced knowledge of the syndrome. And by 1980 analysts were clearly able to differentiate Type 1 diabetes from Type 2 in all their variations. Type 1 strikes females three times as often as males, usually those with a frail genetic matrix. It was insulin dependent diabetes, and usually condemned the sufferer to a shortened life (no more than one's fifties).

Type 2 Diabetes Mellitus was a different creature. Type 2 (that often beset men in their forties) was at least manageable and would still allow me to lead a normal life, provided I followed certain lifestyle modifications. Even if I was insulin dependent (which it turned out that I was), by 1980 they had come up with excellent ways both to redefine it and treat it accordingly

The irony of all this was that it all came about at one of the busiest and most productive times of my life. And it seemed to come almost as a warning to me to slow down the pace a bit.

I didn't of course, but what I did do was start to be more conscientious about my lifestyle choices. Not surprisingly, Bonnie turned out to be my strongest ally in this area. Being an exquisite cuisiniére of comfort foods, my lovely wife soon showed a renewed dedication to keeping me away from my love of desserts and started making sure I kept my food choices healthy.

My physicians also provided me with a strict regimen of daily injections of insulin—one rapid-assimilation shot in the morning, one slow time-release shot at night. It is a regimen I follow in varying degrees to this day, revisiting my dosage levels about twice a year to make sure they are on track.

When I look back upon it, I feel doubly blessed to have had this villain imbedded in our genetic matrix popped up precisely when it did. Had it come to me earlier in my life, I would have been forced to face the mystery of maltreatment, primitive remedies and overdoses of insulin that plagued diabetes sufferers literally for generations. Noting its ravages on the women in our family made me appreciate, at last, the utter Hell they must have endured.

I appreciate even more that it didn't slip through undetected. Undiagnosed Type 2 Diabetes can be a killer especially among men in their fifties because it often presages the onslaught of a coronary. And the younger you are when you get your first heart attack, the less prone you are to survive. Knowing my nature and the fact that I have shown no regard for my body where my work was concerned, I might easily have been ambushed by this; and the offshoots might have been dire. Blindness and deafness often accompany this syndrome if it is not caught in time. So, in more ways than I can count, my timing could not have been better.

As it was, I was faced with my options. And, as was often the case, I proved to be the "crash test dummy" for the other men in my family. Admonished by my own experience, two of my brothers were alerted to this dark demon of diabetic DNA lurking in our family tree, got tested and eventually treated for similar levels of Type 2, as did one of my nephews.

For my part, I have had Bonnie at my side to be my conscience and keep me away from all the bad foods that I love, and to be the constant monitor for those cautious moments of friendly policing that I too often forget. She is the lion at the gate. And I rely on that strength.

This pause to reboot and adjust to my newfound health challenge, in another valuable way, set the stage for the rest of my life. It carried a warning sign: *Work hard. But always be mindful that you are here for a purpose. Balance! Life is all about balance.* (My hardest lesson to learn.)

What it allowed me to do at the time was to learn how to pace myself. I could still accomplish all I needed to do. Only now, I had to do it with an eye cocked toward my health. Too many of my loved ones had died before their time. And I wasn't ready to join them; at least not yet.

My best work still lay ahead of me…as well as the best part of my life. This turned out to be a pause for repairs. And I thanked God for that!

CHAPTER 31

FROZEN GROUND: THE POLITICS OF ENERGY

If it has possibly escaped your notice by now (and occasionally I lose sight of the fact myself), soil scientists these days are almost always called in to clean up someone else's mess. Especially in that ecologically complicated era from the 1970s through the 1990s, we at CRREL had become troubleshooters, often those of last resort. It had become a part of our evolution, and perhaps that was not necessarily a bad thing.

As pivotal players among environmental sciences, soil scientists and biogeochemistry, we were originally focused on examining the mineral (nutritional) content in the native lands of developing nations such as India, Indonesia and Malaysia. In the beginning this was undertaken as a means of defining the levels of depletion and providing the missing nutrition. So, in our way, we began more or less as nutritionists for the soil, making sure that Mother Earth got her fair share of trace minerals such as copper, iron, cadmium and zinc, etc. to be a healthy growth medium for the people who relied upon it.

It seemed such a simple and noble task in the beginning. Soon enough, however, we all came to realize that the politics of energy would have their say. Oil, heavy industry and factory farming, fossil fuels and the machinery of war—all these and more would flood the good Earth with heavy metal cocktails and chemical pollutants in such sophisticated combinations that it would take a genius to identify, analyze and resolve them all. (Several, in fact, and that's where we are very pointedly called back into the fray.)

With the work we had done at CRREL by the early 1980s, we were well on our way to proving our Wastewater Management and Renovation Techniques to be the most effective means of converting what amounted to a potential plague of heavy-metal contaminants into environmental "good guys" and effective, healthy growth media for both water storage and healthy global crop production. (To be sure, we hadn't won the battle yet, but people were listening. In fact, municipalities from California to Australia, and from Israel to China were starting to take our renovation technology to heart and applying it.)

As far as I was concerned, it was a challenge met and mastered. And it was happening in an area in which I was directly involved—hands-on—and able to make what I hope was a measurable contribution. The next level of challenges we ran into were a bit more convoluted and fraught with energy politics. But because they had broad-spectrum global ramifications we had to apply all the integrity at our disposal.

Even as late as the 1980s, CRREL still had several primary areas of focus that crossed over into cold weather tracking and monitoring USSR frozen ground military positions, and its effect on heavy movement was always a primary concern. Formed out of a set of Cold-War military strategies in 1961, CRREL always had one foot in applying science and technology to complex cold-region environments all over the world. By virtue of its own Mission Statement CRREL's primary purpose has always been… "building innovative products which support the warfighter, water resources, environment, infrastructure and homeland security."

In other words, CRREL originally justified its very existence by having government funded logistical support models for an effective cold-weather war machine, primarily in the Arctic and Antarctic. Even the civil and industrial developments we made in the late 1970s and early 1980s were looked upon as peripheral considerations. CRREL was still primarily a wing of the US Army Corps of Engineers. And ACE's major areas of concentration were still mobility and adaptability to the native environment and using "frozen ground" and *permafrost* technologies as tactical enhancements for cold weather maneuvers and "defense."

It has always been one of the ironies of the machinery of war that the technology created from it actually accelerates peacetime advances in science, electronics, communications and engineering. So, in spite of itself, *war actually puts the technology of peace on a fast track.* That usually happens because some "science nerd" somewhere along the way looks at all this paraphernalia and says to himself: "We can make this work in any number of other ways."

That's what occurred to me when I took a good look at the permafrost technologies that had originally been set up for tactical military initiatives and defense strategies during World War II. This technology carried all the way through 1961 (Cold War Era) when CRREL was officially formed, and it played nicely into our remedial mindset when CRREL also became charged with the responsibility for environmental clean ups, seemingly all over the world. These "reclamation tasks" included all military installations and operations—equal to a 110% paradigm of restoration. In other words, the military and ACE were required to "leave it better than you found it," by the DOE and the newly formed EPA. So that job inevitably fell to us.

Fortunately, I am one of those who come from the mindset that embraces the Law of Pure Potential, and that's why so many of our applications were given the funding they needed.

Frankly, it came about through nothing that I was able to do personally to advance the cause of our research in the field. Instead, it came through the now infamous Superfund enacted into law by Congress in 1980. I say "infamous" because it was originally set up through something called the Comprehensive Environmental Response, Compensation, and Liability Act (CERCLA) that empowered the EPA with $1.4 Billion to clean up any *hazardous waste sites* they mandated to be necessary.

Now, something that in today's paradigm would seem like a drop in the bucket, this "fund" had come as an emergency measure to deal with a spate of environmental crises such as the one at Love Canal in New York in 1979 (that almost destroyed Niagara Falls), and the "Valley of the Drums" in Kentucky in 1978 that had polluted the Northeastern part of that state for nearly 20 years.

It seems ironic now, but the Superfund turned out to be anything but that. In fact, only $48 million of the $1.4 Billion allocated was ever used. Most of that was parked on the side of the road by the Reagan Administration in 1981, because it hired a counter-environmentalist Secretary of the Interior named James G. Watt whose sense of the ecology was giving Carte Blanche to the oil companies and shooting deer from helicopters.*

What Watt and the Industrialists of the 1980s managed was a burden of proof to determine hazardous wastes that was so obstructive it would have taken the EPA years to justify even one case. The good news for us was that about all the EPA could do for about the next 8 years was conduct a list of "feasibility studies" for waste pollution and waste remediation and removal. And that once again, to me, was a casebook example that Crisis Creates Opportunity.

In 1984, I was able to submit a proposal to the EPA Superfund for which we were able to secure a $365,000 budget to conduct a Frozen ground pilot study through CRREL, one that would show that pollution solution theories might indeed become axiomatic.

To accomplish this, I created a "theoretical" model using septic tanks on adjacent land, and working with a graduate student at Dartmouth named John Sullivan, we collectively examined the metals, movement and diffusion of contaminants through the frozen soil.

I was stepping on a territorial minefield here (every pun intended), because CRREL, though they allowed the study, didn't really care about the environment in general. They were only interested in the work we were doing insofar as it allowed us to come out with ready solutions for controlling spilloffs from ACE studies on ordinance, military fuels and the toxic runoffs they experi-

---

* James G. Watt was adjudged to be the worst Secretary of the Interior ever to act as "steward" for the US Environment, primarily known for "de-authorizing" national parks and eliminated the Land and Water conservation fund. He resisted major reparations set aside for the EPA superfund. And he was later, in 1995, indicted on 25 counts of obstruction of justice, perjury and influence peddling as a lobbyist. He was sentenced to five years probation on a misdemeanor for withholding documents, forced to pay $5000 in reparations and 500 hours of community service.

enced with TNT, cannon fire and residual contaminants from aircraft such as jet fuel and diesel.

In a way, this put them at odds with the EPA, and set them off into something of a series of turf wars, because CRREL wanted validation from the Superfund study but not EPA oversight and "snooping" around because of it.

What this enabled us to do was extend our "frozen ground" bioremediation studies into the next step—a broader test area at a nearby plant at Oak Ridge and finally to Alaska where the major portion of the study would be undertaken in 1987. The first aspect of the study on ground freezing technology was to examine the theoretical cost of energy infusion as an aspect of restricting contaminant movement, and it was determined that energy costs cannot be an object in consideration of the technology of containment.

Historically, it is important to understand isolation for hazardous waste removal went all the way back into the 1880s, when frozen ground was used to section off old mineshafts (gold, silver and even coal), to keep the fuels used during the mining process from sieving over into the native groundwater and polluting it. Later, up to the middle of the 20th century, weather and civil engineering construction were able to form a frozen barrier around military explosives, isolating them and then applying a normal "freeze-thaw" cycle methods to extract any contaminants.

This was all weather reliant, and worked best in places with long winters and long days (or nights), which made Alaska the perfect area of study.

What I was proposing in general was something far more radical because I believed that we could use "frozen ground" technologies 365 days a year, at any time to contain pollution and isolate biohazards. And we could use a logical, well-regimented three-step process to accomplish a rather complicated objective: *1) ground-freezing; 2) decontamination; 3) removal and extraction through the addition of chelation techniques.* Let me explain.

Ground freezing can be used to stop any movement of contaminants from any source—fossil fuels, runoffs from explosives, wastewater and sludge, and industrial runoffs—by forming a permafrost barrier around the contamination. This *first generation application* is done primarily by the insertion of freezing stations or vertical barriers to lock off the pollution and hold it in place. (Originally this was done as a seasonal freeze-thaw cycle to gather the contamination and remove it.) Based on the theoretical model we came up with in 1984, if you want to freeze and contain the contaminant, your primary objective should be to drill as close to the source as possible. With modern technology, you can insert freezing applications—walls, insertion pipes or stations—any time of year. And you can hold the contamination in place indefinitely.**

Decontamination is virtually a matter of proximity. And what I mean by that is that, once you've isolated the contaminant, you use the freeze barriers to move it away from the affected area, forming a kind of wall that shoves the frozen toxins very much like a snowplow or an earth shovel would do. So the closer the freeze technologies are to the contamination, the easier it is to control, move and manipulate.

The Final Step of effective *Removal and Extraction* is still a matter of some debate. For me, it is really a closed-case because our years of constant research and application point to *Chelation* as the most effective means of isolation and removal. By chemical composition, chelates are agents that remove heavy metals and contaminants at the source—especially lead, mercury and the more dangerous elements. They do this by causing them to become "soluble," so they can be blended into one form, brought to the surface and extracted at the source.

In our tests and findings, an element called *EDTA* as the best means of removal. EDTA *(ethylene diaminetetra acetic acid)* is also present as a detoxify-

---

** One of the technologies I have always insisted could be introduced is a "frozen failsafe" has been pre-established freeze barriers in areas where known pollutants are present. These include fossil fuels, industrial runoffs and military ordinance. When they occur, the best practice is to have the freeze barriers already in place. That way they can be turned on virtually like a "freeze faucet" whenever a pollution event should occur. Properly applied, this could throw up a freeze barrier in a matter of hours rather than wait days or weeks for the system to be set in operation. In my opinion, it should be a required precaution at every Nuclear Power Plant in the world. If the Nuclear Reactors at Fukushima had this kind of system already in place, this international disaster that has quite possibly contaminated the entire Pacific Ocean, might never have taken place.

ing agent in some foods as Disodium EDTA (look for it in pickles and mayonnaise). And though its use in the food chain is occasionally challenged, its application for industrial uses is a significant benefit. What the EDTA is particularly adept at is forming a cup for the contaminants to catch the metal, make it soluble and pump it above ground, ultimately running sulfides as isolates to remove it.

This "frozen ground" methodology may sound simple and basic enough when you see it. In reality it took over ten years of research, of trial and error a team of brilliant bio-geochemists and soil scientists to come up with these answers. They're still not universally accepted, even though we now have literature and tests to prove that it does.***

Of course, so much of the work we had to do with CRREL and as soil scientists involved reclamation and remediation rather than immediate on-site emergency applications. For that reason, from 1986 throughout the rest of the 1980s and the 1990s, I was able to open Alaska as a major field of research and repair for CRREL (even establishing a branch there for commercial purposes).

In locations all over Alaska during the late 1980s, I was called in on at least three separate occasions: first to provide solution models for with residual spillovers from explosives like TNT and fuel from military installations in Alaska

---

*** Very often my work enabled me to team up with some of the finest minds in biogeochemistry, including my working associate Steve Grant. And it has often been my privilege as well as my position that has enabled me to champion their works and add my contributions to their publication. Such was the case with these choice books and key studies:

Grant, S. A. and I. K. Iskandar (Eds.). *Models for Cold Regions Contaminant Hydrology: Current Use and Future Needs,* Ann Arbor Press. Ann Arbor, MI. 1999.

Ayorinde, O.A., L.B. Perry and I.K. Iskandar. 1989. "Use of Innovative Freezing Technique for In-Situ Treatment of Contaminated Soils in Proceedings." 3rd International Conference on New Frontiers for Hazardous Waste Management. September 10-13, 1989, Pittsburgh, PA pp. 487-498.

Chamberlain, E., I.K. Iskandar and S.E. Hunsicker. 1990. Effect of Freeze-Thaw Cycles on the Permeability and Macrostructure of Soils. Proceedings International Symposium on Frozen Soil Impacts on Agricultural, Range, and Forest Lands, K.R. Cooley (ed.), March 21-22, 1990, Spokane, WA, pp. 145-155.

from World War II; next to deal with the jet fuel and diesel runoffs from the years of air traffic and jets at the (Ted Stevens) Anchorage International Airport and nearby Elmendorf Air Base; and finally another to consult on the spill-off recovery and reparations made at Prudhoe Bay, home to the largest oil fields in all of North America. †

In both instances, if you were going to apply any kind of ground freezing technology, you first had to set up to remove oil and clean the soil, and then go in slowly to see how far the oil permeated the frozen soil. In these cases, we didn't move much oil. Instead we used permafrost as storage for the oil, moved it along for decontamination and then brought it to the surface for clean up. Having actual soil is a good idea, because the cost of doing that is a fraction of cleaning the site. That way we could protect the permafrost, and help prove the theory that ground freezing can provide the perfect barrier to preventing contamination. So we were able to use hands-on experience to prove our test points.

By that time in the late 1980s and 1990s I was willing and able to turn at least some of these hands-on tests and studies, while I focused more on setting up conferences to advance both our knowledge and our work in frozen ground technology. This became at least a part of my new job description as Branch Chief of Geosciences. At least one of the reasons we were able both to secure funding and to receive ACE's blessing to conduct all these frozen ground studies in the 1980s and 1990s had at least something to do with the fact that, in 1986 I had been appointed Branch Chief for Geochemical Sciences at CRREL.

I hadn't originally intended to do so, but there were four other applicants, and the associate with whom I had some interaction seemed to be in the lead to get the job. He was a decent fellow on his own but simply couldn't handle the pressure, and seemed to become both abusive and self-sabotaging whenever he was confronted with a difficult challenge.

---

† Recalling my first trip to Prudhoe Bay in 1986, I arrived in the dead of winter in -40º (Fahrenheit) weather and thought it was about the most miserable three days I had ever spent anywhere. Returning to Anchorage in August 1989, I was totally enchanted by its beauty and actually looked forward to my return trips to Alaska, as long as they took place in the summer.

Since I seemed to thrive on pressure and had the support of my co-workers going in, I applied for the position and got it; and I think everyone was relieved. (I certainly know I was.)

What my new role as Branch Chief enabled me to do was to gain leverage and set up conferences and "exchange" concepts with our old Cold War archrivals the Russians, especially in the early 1990s. By then, with their new West-friendly incarnation under the leadership of Boris Yeltsin, we had an open gate to information that I might never have been able to unlock in earlier years.

What I learned through a number of global gatherings and interactions in venues ranging from Fairbanks, Alaska to London was the fact that, dating back to their days as the USSR, the Russians and the Japanese were so far ahead of us where "frozen ground" technology was concerned that the best thing I could do was to become an avid student of what they had developed up to that time in terms of the troika of *freezing, decontamination, and (chelation) removal.*

My dual role as both scientist and organizer brought me a lot of lessons along with the responsibility: And Rule # 1 is that you never stop learning. You never cease meeting people in your field who can teach you things you never might have otherwise considered; and in science, as well as business, *networking* often proves to be an even more valuable commodity than information.

So often, in science as in business, it isn't always what you know. It is who you know…and what you can learn from them. That is the business of smart science. That is the science of smart business.

CHAPTER 32

THE SCIENCE OF SMART BUSINESS

It is, at this point, safe to say that science is often an art. By the same token I have to say that business—more often than not—turns out to be a science. Permit me to illustrate.

Very often in scientific research and experimentation, you build a model or a concept that, though based on a set of facts, often becomes an act of faith, a belief in an outcome that you may have to find some very creative ways to prove is a reality. It has become a common truism to note that most scientific theory "begins as heresy and ends as superstition." What that means is— even though you are doing the best you can with the knowledge, the preset formulations, the paradigms and the mathematical models you have set up— you are still facing a universe of uncertainties.

In business, it is ironically a little different. There are some rules that apply. They're called guidelines. Some are even laws driven by a set of rules. Play by the rules, and you have a very good chance of winning. Try to reinvent them, and you will very often find yourself in trouble, if not broke. It is often a matter of "dollars and sense," and just some basic math. And, to be sure, good approaches to smart business can be very scientific.

In many ways the rules can be ironclad and covered by some very basic philosophy, beginning with that Holy Trinity of real estate investment: "Location! Location! Location!"

From that point, bank on your essentials: "Buy low. Sell high. Never lease when you can buy. Buy things that appreciate. Lease that which depreciates. Your own time is your most valuable commodity. You will never get rich working for someone else. And there are two sure-fire ways to avoid dying broke: Be born rich. Or have multiple streams of income."

There are other maxims that also hold true: Never invest in things you know nothing about. Never take investment tips from taxi drivers. If any commercial venture seems too good to be true, it usually is. There is no such thing as an easy buck. High return on interest usually involves high risk. Never pay someone else for doing something you can do yourself. The biggest mistake most people make is not investing in themselves. And finally, windows of opportunity only open if you've done your homework first.

Every one of the things I have just laid down in the two previous paragraphs are true. And virtually all of them have exceptions that prove the rule. The issue comes with knowing when to apply the science and recognizing the science of smart business for what it is.

Up to this point, I have made no bones about my penchant for hard work. It is a matter of pride with me. I've always worked harder than anyone else. It just took me a very long time to learn how to work "smarter" than most people. And success is directly proportionate to an intelligent application of skill.

There are a few very fortunate people who grow up in financially informed families that teach their children sound economics and the essential strengths of capitalism virtually from cradle to grave. They are exposed to the fearless culture of the entrepreneur and get to approach the building of income and amassing of portfolios with the same expectation, as one would have of running fast or mastering multiplication tables.

Most of us, I'm sad to say, are trained in the tradition of what can only be accurately labeled as "corporate servitude." Whether we work for the government or a large conglomerate, we are taught to give 78% of our waking time to an employer who pays us 12% or less of what we're really worth, and we do it in hopes of trading long term job security for a couple of years of peace and prosperity at the end of our lives.

I grew up in that world. I saw my father slave away in a low paying government job, doing things he generally disliked while remaining terrified of his own dreams of personal business independence. In some small way, I have finally come to understand Karam Iskandar's gambling addiction, because those few evenings out on the edge were probably the only times in his life when he truly felt alive, in charge and with some smattering of hope for breaking free—for financial independence and a new window to the future

My resort through most of my young manhood was the opposite of that. I worked "two jobs" until I was in my sixties—my main job at CRREL, and my other in real estate, property management and maintenance. And I did so because, even early in my career, I found that investing in myself was one of the few things I could do with a high degree of certainty. Ever aware that my second jobs were demeaning minimum wage positions at the tail end of the workforce, I still felt that there was still come dignity in them. And it often meant the difference between having a meal or not.

Since I had originally been schooled in the scientific tradition, my most viable sources of reward often came in the form of scholarships or research grants, which created a culture of dependency on benefactors—rich corporations, the government or the military. So it took some shaking-up for me to come out of that trickle-down bureaucratic mindset.

I got my first taste of it—that heady rush of financial independence—in 1976 when I bought my first home. I had been renting all my life. And the first thing anyone who has rented will tell you is that it's loser's game where you see your hard-earned money go straight down the rabbit hole into someone else's pocket, while you continue to scratch to make ends meet.

I also remember when my parents were finally able to become co-owners of the house on 19 Refaat Street how much it helped change our family dynamic. For the first time in my young life, I could see how it affected us all—with a sense of empowerment, the pride of ownership, and a control over one's own destiny. I could feel the tensions ease at last, and I wanted to extend that experience.

When I bought my first property in 1976, I quickly came to realize that you could never make great money working for the government. But job stability combined with decent credit can at least get you to a place where you can borrow money to make a down payment.

After I purchased our first house on Kinne Street in 1976, I didn't know the first thing about leveraging, but recognized immediately the need to learn. So I did. I borrowed the $27,000 with a mortgage loan to do it, making sure that I would repay that money and more once it had matured. What I didn't want to get caught doing was taking out a long 30 year loan on the property, because anyone who runs the math will tell you that anytime you get a 30 year mortgage, you are merely paying interest for the first 15 years. So, the shorter the duration, the better.

Within the same year, I was literally able to pick up a property at what amounted to a "fire sale" price. Shortly before I discovered the property, there had been a five-unit Apartment house in an old converted red barn that had undergone some fire damage.* Because it had, the entire five-unit dwelling was going on the market for less than $38,000. I got a loan, brought the cost of total repairs in for about $10,000, rented all five units, and went on my way to buy another property with the money I had made on the balance.

Along the way I found out just what valuable teachers real estate, purchasing, management and development can be. And in this case, they taught me three things very quickly: First, you can save a ton of money in real estate by putting in the sweat equity yourself. By doing the rebuilding, repairs and maintenance on my own rather than paying subcontractors, I saved enormous amounts of money and applied it toward the purchase of my next property.

---

* *Barn conversions*—converting "classic old barns" into living dwellings had become phenomenally popular in the 1970s, especially in New England. They were often very attractive on the outside while being rife with structural problems and safety issues. Nonetheless the combination of aesthetics and trendy living quarters allowed them to be easily converted and inexpensively repaired for someone with the right kind of understanding of perceived rental value versus market trends.

Second, it is best to purchase dwellings of at least five units. It provides benefits for reasons of ownership, income derived from rentals, and taxes and transfers at the sale.

Third, I discovered one sly investment secret even more emphatically: You make money when you buy the property in question rather than when you sell it. So buying a house, an office building or an apartment complex at "the right price" and maintaining it yourself eliminates costly contractor obligations and enables you to use your "sweat equity" to build in more profit once you put it on the market.

I've never shied away from putting a little elbow grease into any project I decided to take on. So, I decided to eliminate yet another link in the chain of middlemen and superfluous fees. And, sooner rather than later, the logic struck me: Why pay brokers and real-estate agents commissions that you can keep for yourself? Real estate brokering is a tough way to make a living to be sure. And I don't wish to deprive anyone else of his or her income. But what I was finding whenever I went to negotiate a property was that I was much better at negotiation when I dealt directly with the owners.

Real estate agents were so often desperate to make the deal that they would all-too-frequently try to pressure you into accepting conditions that just didn't work. Then there was the "pile-on" factor that seemed to come with every transaction, ones that required you to go through a broker and pay the 6% or 7% commission that you might have kept to yourself.

So, my logic was this: If I became a real estate broker, I could legally carry out my own transactions, negotiate my own terms, and pocket the commissions I would otherwise have to pay to two or even three other people during the course of a probe, purchase and close. After running the math, it didn't take long at all for me to decide to become an "agent on my own behalf," and I went out and got my real estate license.

This meant going to night studies and taking a very difficult set of exams even to achieve that license. But after a couple of years, I was buying, refurbishing and (occasionally) reselling at least two to four individual properties a year. This began in the late 1970s and accelerated through the 1980s, ultimately achieving exponential expansion in the 1990s and over into the New Millennium. If soil science and geosciences fed my soul, real estate, property management and the pride of ownership fed my finances—my family, my future stability and my pride in community in ways I never imagined possible.

To be sure, not all my ventures into "multiple streams of income," turned out to be profitable. In the mid 1980s, although I did seem to have the rhythm and psychology of real estate pretty much in tow, I was not entirely immune to making mistakes…particularly when I decided to go off the game I was winning and jump onto the momentum train called the New York Stock Exchange.

Mind you this was the late 1980s. After a nearly 20 year drought that had run from the mid 1960s to 1982, Wall Street was finally enjoying the longest sustained bull market in the 20th century—five years of unprecedented growth and expansion. And I was doing so well in the synergy of my double career as scientist-entrepreneur, I decided to go in for the kill and get some of that "easy money" that was floating around in stocks.

Five lessons, right out of the gate, one of which I totally ignored:

1) Do not invest in something—a company or an industry—unless you know everything about it.
2) Never try to invest on momentum alone without first checking on Galton's Second Law, "that everything reverts to the mean."
3) Don't buy on margin.
4) If you're going to go into long-term investing, bonds are better than stocks. And finally,
5) is the lesson on greed: or in the old Wall Street cliché, "Bulls make a little. Bears make a little. Hogs get slaughtered."

By 1985, I had been doing so well in real estate investing and acquisitions that I took a nice middle-six figure profit and rolled it over into the stock market, buying what I considered to be some very solid stocks. My thinking at the time was this: "What could be a safer investment than a bank? So, I know I can't own a bank. But I can own part of a bank by buying stocks in that bank."

Over a protracted period of time I bought shares in banks—not just one bank but half a dozen banks, and not just any banks but the banks in my own backyard. I bought Valley Bank, Hanover Bank, Marble Bank, and Bank of New Hampshire as well as some others, mostly banks and people I knew and relationships I trusted.

Then on October 19, 1987, The Stock Market had a "correction." The Dow Jones Industrial sank about 508 points and lost 22.6% of its total value in one day. (In today's market, that would be equivalent to a drop of about 4300 points.) Virtually all stocks, especially bank stocks, tanked with it. So my "solid investments" cost me dearly. (As an example: One bank stock I bought at $27 a share in 1986 went down to $3 by 1988.)

The bottom line was that this "crash" turned out to be real, and the global markets didn't entirely recover until two years later in 1989. And yet it was in that same year when the financial markets really bottomed out with the infamous Lincoln Savings and Loan and American Continental financial scandals whose $6 billion in losses not only ended the political careers of some key American Senators but also trashed almost every financial stock in the nation—including such icons as Chase Manhattan, Wells-Fargo and Citibank.

By that time my investments in financial institutions ended up costing me about half-a-million dollars. And though I allowed some of the stocks to recover, I was forced to write off a large amount of the investment.**

Meanwhile, I had gone on to learn other lessons. One of the first ones was not to panic. When stocks go down, either buy them when they are or give them the opportunity to regain their balance.

** One of the biggest mistakes I made was in believing that losses on businesses investments had the same tax-loss implications. They don't. And write-offs for stocks that tank are the worst. In fact, your total write-off for money lost on stocks or bonds is a paltry $3,000 a year—not per stock, per year!

A second lesson was to go back to what I knew, to "Sticks and Bricks" as the people in real world of real estate call it. I started focusing more on real estate and property management and rebuilt the fortune I had amassed.

Perhaps third among them came with the realization that medium and long term investment bonds, provided they're not junk bonds, are far more stable than stocks in that they mature at their redemption value and repay the investor all the interest accrued over the 5 or 10 or 20 years inclusive in their redemption.

Following these guidelines, over the long run, has enabled me to make all the money I lost and much more. And it taught me a valuable lesson. There are some forms of investment, such as following the madness of the Wall Street Mob, that are no better than shooting craps or trying to hit "the numbers" on a roulette table (something I think my father would have understood).

Upon reflection, the 1980s were some of the most active and productive years of my life. I was now following two careers, learning all the ups and downs of intelligent acquisition, income building and tax benefits. Mind you, I was managing this sort of double life during the late 1970s and early 1980s when I was also utterly immersed in my research and some very landmark studies at CRREL. So, as usual, there wasn't much time left for a family life. Frankly at the time, I was so utterly consumed by this great professional one-two punch in my life, that it had possibly become another kind of addiction.

Add to this the fact that in 1981 my wife gave birth to our daughter Amy, and my challenges in trying play substitute dad to Bonnie's two sons Bruce and Tony, and I was made very aware that I had to make optimum use of my time in all three areas.

In a unique convergence of the dual family dynamic, my son George, who had just turned ten, decided "to come and live with my dad." So it soon became one of the busiest and most joyous times in my life. Yes, of course, it put more pressure on us, but it was the kind that I welcomed. I had an expanded and somewhat mixed family now. For some it had been a matter of choice; and I was more than ready to rise to the occasion.

Given the fact that I have always had an abiding faith in a kind of loving God, I came to have three priorities later in my life: my work with the government, my career as a real estate entrepreneur, and my devotion to family.

Upon reflection, I think I truly believed that the best way of taking care of all those entrusted to my care was to work harder, provide more financial security, and be a source of stability—to not one but two families. Not only would my new baby girl and my two-step sons be well provided for, I could now also personally and financially take care of my daughter Niveen and my son George in ways never available to any of us before.

Given my impoverished upbringing and having spent much of my "student life" struggling with financial challenges, I wanted to be the best provider I could be, to make sure that my daughters and sons had the best housing, clothing and educational opportunities money could buy. What I may have lost sight of somewhere along the way was that emotional support is an intangible that many of us unwittingly ignore.

Frankly Bonnie was so nurturing and so supportive of us all, I think I relied on her to provide the emotional accessibility that I felt I lacked. My Egyptian social traditions probably didn't fit here. Egyptian men in particular were not expected to show either vulnerability or affection. (But this was no longer Egypt; this was America. The parental dynamics were different here, and I should have taken the time to look at that.)

I don't think I ever slept much in those days. I would work long hours at my Branch Chief's job at CRREL. I would spend all my weekend time working my real-estate business and my investments. And I would devote what time I had left to property management, construction, trouble-shooting and dealing with tenant needs.

In fact, I spent almost all of my free time in those days immersed in all aspects of property management—ground repairs, recycling and trash collection at my various properties. I think I tried to use that time to bond with my sons, and later on with my daughter Amy. Once she came of age, Amy actually enjoyed our little "adventures in recycling," and it became a game for us, one that would reward her for redeeming bottles and containers.

Amy always showed a knack for business and the games that came out of our little weekend work projects. I don't think either Tony or Bruce ever quite got the hang of it. Somehow, I tended to come off as that crazy Egyptian stepdad who often came home late at night and tried to put them to work. So their "games" became ones of escape and evasion, and in time they got quite good at it. (In fact later, after they had grown up and we had Thanksgiving or family gatherings, the kids reminded each other of the clever ways they used to avoid work details and helping me with "the business.")

That in itself was something of a skill, and I think if I had understood how they felt at the time, I might have been able to channel my passion for success a bit more effectively. I could have been a better parent in the bargain. One can always strive to do better.

As it is, success can be a terrible master in that its rewards and demands are constant and relentless. You get to taste the prosperity it provides just as you come to understand that, unless you work hard to maintain it, it can vanish in the same way that it came.

In the 1980s and early 1990s I had been able to achieve many of the levels of success and accomplishment I had always hoped for. And for a brief time I believed, somewhat incompletely, that I had mastered The Game. What I came to find out, and soon, was that I was just beginning to learn…and that there are more tricks to acquisition and profit than I ever could have imagined.

CHAPTER 33

THE 'TAX MAN' AND THE 1031 EXCHANGE

I'd like to begin this chapter with a personal note of thanks to the Internal Revenue Service. If it hadn't been for your constant assaults on capitalism, some of us might have never gotten as creative as we had to and used your own codes as a fulcrum to gain financial leverage.

    As in all facets of business and finance, one has to understand the game clearly. And the rules by intention are this: The IRS wants to extract as much money from private business, industry and property as it can without entirely depleting the source. To adapt an old fairy tale metaphor: "It doesn't want to kill the Goose that laid the Golden Egg."

    About 85 years ago (actually during the height of The Great Depression of the 1930s), the IRS established and refined what was then known as Section 1031 of the Internal Revenue Code. According to Section §1031 of the code: "The exchange of certain types of 'like-kind' property may defer the recognition of *Capital Gains* (or losses) due upon sale, and hence defer any *Capital Gains Taxes* otherwise due." Bottom line translation: they were offering a way of making real property transactions without taxing you to death when you did it.

    The important hook here is the term, "Like-kind." Because the dictates of the code specifically state the following: "The properties exchanged must be of *like-kind, i.e.,* of the same nature or character, even if they differ in grade or quality."

Personal properties of a similar nature are *like-kind properties*. However, you have to be careful in how you do it. There is a caveat woven into the fine print of the 1031, specifying that personal property used predominantly in the United States and personal property used predominantly elsewhere *are not* considered like-kind properties.

This codicil was set up for two distinct purposes. One was to emphasize "tangible physical property assets" to eliminate any finagling where stock, bond or other "paper transactions" were concerned. The second was to keep all tax benefits regarding real property inside the 50 United States and its territories, such as Puerto Rico and the Virgin Islands, and not see all sorts of valuable US real estate go slipping off into easy-access foreign ownership. Any like-kind property transactions inside the United States and its territories may benefit greatly from tax deferrals, if they fit into a certain time-date framework.

(As tax codes are constantly being amended, this new structure was refined and updated in 1954 and has remained in place to this very moment. This was a reorganization of all of their laws affecting exchanges that helped structure the modern framework of like-kind property exchanges.)

So the real advantage of the 1031 Exchange is for real-estate ownership, sale and repurchase. Some also refer to this real estate/income tax maneuver as a "tax-free" exchange of real estate property. In reality, this sale of one property and purchase of another is only tax-deferred.

This means that the capital gains that one would normally have to pay income tax on from the sale of a property is delayed for a predetermined time if the sale qualifies inside the framework of the 1031 Code. In such an exchange, the property that is being sold is known as the "relinquished property," and the property being purchased is called the "replacement property." And the replacement purchase had to be made *within a specific window* or else the tax benefit would be lost.

Although I had been a licensed real estate broker for nearly two decades, I wasn't familiar with the phenomenal tax breaks or benefits of these like-kind exchanges until I stumbled upon it in 1999.

At the time, in August of 1999, I was in the process of selling a large apartment building known as Stone Farm when I was finally introduced to the concept. Stone Farm was a complex of 36 apartments that were built condominium style as three detached buildings. Each building had twelve units, each with three floors of four apartments. Each apartment was a two-bedroom unit. In addition to the three buildings I had a building permit in place for an additional 104 units.*

I had bought this property from the Federal Deposit Insurance Company (FDIC) in 1992 for about three-quarters of a million dollars, and now stood to benefit from a sale price that was somewhere around four times that amount.

This was due at least in large part to improvements I had made in the property over the years. I had heated the buildings with safe, inexpensive liquid propane, had weatherproofed them all, and added new roofing and insulation. I had also gotten permits to put up 104 units on the land around it, but was confronted with the fact that the construction would be taking place on a shelf of solid limestone—some of the most difficult and costly sub-strata on planet earth —a construction nightmare. Faced with an additional $1 million in out-of-pocket excavation costs, I decided not to get into the one area of real estate I had never been comfortable with in the first place—building and construction. So I decided to do the sensible thing and sell.

However, knowing that the apartments were highly appreciated, and that I stood to pay a huge capital gain tax on the sale of them, I was trying to find some way legally to get around the $500 K of income due to go down the IRS capital gains rabbit hole, unless I came up with something "creative." That *creative solution* came a short time later, when two other real estate brokers

---

* In a bittersweet bit of nostalgia, I flash back to the time when I first came to Hanover, 1975, Stone Farm had just been completed and was the "prestige address" for young families and upwardly mobile singles in the area. I had visited the facility, and wanted very much to move there. Unfortunately, at the time I was still reeling from my alimony and child-support payments, and I thought the monthly rent plus expenses was a little too rich for my blood. And here I was in 1989, fourteen years later, buying the whole complex and having no problem writing the check.

with knowledge of the mechanics apprised me of the benefits along with the belief that my property would fit nicely within the §1031 framework for this sale, provided that I could find a "like-kind" exchange within the allowable time-frame.

I immediately began studying the concept and investigated the process to evaluate the pros and cons of this kind of transaction only to find that it had been used successfully in virtually thousands of transactions inside the US every single year. And it certainly fit the parameters of the one I was about to undertake.

My broker friends both recommended a *Qualified Intermediary* (QI) named George Foss in a nearby community just north of Hanover in New Hampshire.

The Qualified Intermediary, or Accommodator, plays a crucial part in the exchange transaction because he or she acts as a middleman (a de facto agent) between the seller and the buyer of the real estate. As such, the QI is responsible for holding in escrow all of the proceeds of the sale of the relinquished property until the purchase can take place for the replacement property.

Since experience had taught me to have an innate distrust of lawyers, real-estate brokers and CPAs, this was absolutely the hardest part of the transaction for me to accept.

Here I was, essentially, handing a stranger close to $3 million dollars! Needless to say the benefits were so obvious that I scheduled an appointment to meet this George Foss, and within a half hour of meeting with him both my wife and I were convinced of his honesty and professionalism. This was a major turning point in my investment future and was a major contributing factor toward my being able to build wealth and financial security.

In nuts and bolts terms, each individual transaction has a specific process that must be followed exactly in order to qualify for the 1031 tax deferral. The like-kind exchange is basically this: You sign an agreement to sell the relinquished property. Within 45 days you have to identify a "short-list" of up to three properties with the specified intention in place that one of those three will be purchased and become the replacement property.

In turn, the indicated replacement property must be purchased by the same person(s) or entity that originally held the relinquished property. And the replacement property must be purchased within six months of the sale of the relinquished property.

In the case of the Stone Farm contract, in order to start the process of this first exchange, I needed to find a replacement property since I already had a buyer for Stone Farm.

At that time, as I was approaching retirement, Bonnie and I were considering moving south into a warmer climate. We thought that it might be nice to start buying some property in those states now in order to make things ready for us to transition from New Hampshire. We arranged a visit to a condo that we already owned on Hilton Head Island, SC and on our last afternoon there, we drove by a commercial property for sale with our real estate agent that we both really liked. This property was called the Shipyard Galleria and housed several tenants including a Blockbuster Video, a Greek restaurant, an insurance agency, and a health club.

We purchased the Shipyard Galleria for about two-fifths of the proceeds we received from the sale of Stone Farm. At the same time we bought the Galleria we also purchased an office building in Fort Lauderdale, Florida and a farm house and land in Lebanon for the balance of the exchange and all the tax benefits it encompassed. So, at the end of this first transaction, the total value of the three properties replacing the relinquished Stone Farm came to a bit less than the profits from the Stone Farm like-kind transaction. And all federal and state taxes were deferred.

By 2005, we had sold all the properties from the original 1031 Stone Farm transaction, triggering an entirely new series of like-kind exchanges. In each case, the proceeds were reinvested into new properties. Since prudence and financial advisors have counseled us to be discreet, I will simply suggest that you do the math.

My only suggestion to any of you looking into the potentials of the §1031 Exchange in your state is that you work on multiples of five when you do…and with proportionate tax benefits for each. That is, of course, just part of the story. The rest of it becomes even more fun. And to be sure, we were to run into some incredible like-kind opportunities along the way.

As I mentioned at the beginning of this book, there are times when the Laws of Attraction apply. And soon enough that becomes the First Commandment of Good Business.

CHAPTER 34

A FAMILY AFFAIR: REUNION

Perhaps it is my role as the eldest son, but I have always held onto a strong sense of family. It is both a life-pattern and a personal feeling of responsibility. It is a tradition in most cultures for the "first born" male to take his place as the family patriarch once the father has died. And along with it come all the responsibilities and burdens of expectation that one might expect.

Although I never sought it as such, I feel as if I assumed that role even by the time I'd reached the age of 30 and had permanently moved to the United States. For a score of reasons—some personal, some involving issues of career and academics—I have experienced more of my family joining us on this side of the ocean than those who have stayed in Egypt.

Driven by a half-dozen different motivations, I never doubted my own decisions to make the change. Once I had gotten over here, I realized how right I had been in opting for America as my new chosen home. Not only were the academic opportunities unlimited, so were the career horizons and the lifestyles that came with them.

America always has been a haven of unlimited potential and a place I personally resonated with from the outset. Once I got over here and onto my new status with the University of Wisconsin in the Soil Sciences Department, I was certainly intent on sharing my newfound love of country with the rest of my siblings as well as my own parents. And I was both surprised and disappointed to find that my wife Marcelle, when she joined me in 1970, shared little of my enthusiasm.

I thought it might be different for the rest of my family. At least I hoped it would be. I knew it would present some excellent career opportunities for my brothers if they chose to pursue them. But since they were all much younger than I, it seemed only logical that it would take some time for them to come over and experience this remarkable land of opportunity for themselves.

What I learned early on and underscored later was the fact that any two people will extract entirely different outcomes from the same set of circumstances. The difference is one of motivation. That's what my mother hoped I could instill in my younger brother Atif when she pleaded with me in 1968 to have him come over to America "for a visit."

Atif is the middle brother of my three younger brothers and was born 15 years after I came into this world. So in many ways, I have always functioned as a father figure as well as an older brother to him…and to all my siblings.

When he was eighteen, Atif had finished high school but seemed to show no inclination to go to college or to get a job. Vastly talented and able to charm anyone into anything, he had managed to convince someone else to take his high-school equivalency exams for him. And so, by proxy, he had come out with very good scores as well as being eligible to move forward, provided he chose to do so.

Since Atif was the only one of my brothers who truly got on well with my wife Marcelle, Aida pleaded with me to bring him over to Madison where he was to stay with us for a "trial year" from 1969-1970. Aida's belief was that my work-ethic, my focus and my career connections might possibly inspire him to seek a higher education as well as be challenged by this very dynamic new social environment. But it didn't quite work out that way; quite the contrary.

Although he enjoyed "hanging out" with the family, Atif really didn't get the US at all. He hadn't really learned English well enough take advantage of his social contacts here. So, other than my brother's interactions with Marcelle and me and a few of our Coptic inner-circle, there was really nothing for the young man to do, other than get a menial job in a cultural atmosphere for which he had shown no affinity.

After a few months, Atif asked to go back to Egypt. And though I had tapped into my savings to get him a ticket to America, I warned him that if he went home now I wasn't going to pony up for a return ticket. At the time my little brother was so homesick for Egypt, he insisted this trip would be his last. (And we both believed that it would be. We didn't know it at the time but we could not have been more wrong. He would return about 8 years later and put down roots.)

If anything Saied, Atif's older brother by four years, was entirely focused on coming to America and did so by late 1975. As motivated as I was to achieve academic excellence, Saied was a driven, studious young man who had, by the time he was 22, already earned two Master of Arts degrees, with honors—one in Finance, the other in International Business.

If some other men in the Iskandar family felt little or no affinity for academia, Saied was fixated upon collecting degrees. Saied saw higher education as his first class ticket…to everywhere! And, to this day, he has more degrees than all other men in the Iskandar family, including me.

Following the trail I'd blazed at the University of Wisconsin, Saied arrived there in 1975 very just a few months before I had decided to take the job with CRREL in New Hampshire. But getting him across to this side of the ocean was tricky in the beginning and took some sleight of hand and a little maneuvering. Still we managed to navigate that delicate passageway called immigration, using the complicated machinations of government bureaucracy to work on our behalf.

Ironically, it was not US Immigration that presented the problem for Saied to leave. When Atif had wanted to come over, his "twilight" status inbetween high-school and college still rendered him a ward of my family (a younger relative). So, essentially all I had to do was sign some papers vouching for him while he was in America.

Since Saied already had a degree (or two), he would not be allowed to leave, because once you are graduated from an Egyptian university you were not allowed to leave Egypt until you had fulfilled your obligation for military service. As it was, it took him four years to get his papers, and he had to play a little trick to do it.

Since he already had a Bachelor of Science degree (in Food Technology), he returned to the University of Cairo to get an MS, re-registering for the new curriculum as a sophomore and then transferring on a student passport as a second year science student. This bit of clever academic manipulation enabled Saied to visit me in Madison, first as a vacationing student and then on the same *student transfer visa* that we managed to secure with the US Consulate before he left.

Saied remembers being a bit paranoid about the whole process, even though he'd managed to sidestep the post-graduate restrictions: "I left Egypt with just a handbag because I didn't want to get bogged down. And I didn't mention my degrees so I wouldn't be stopped at the airport in Cairo." In a way, due to the restrictions placed upon graduate students in Egypt, Saied's move to America had more the flavor of a getaway than a visitation. Still, it was an idea whose time had come, and everybody endorsed it.*

As opposed to Atif who was too young and not bilingual when he came, Saied had a good command of English, a great facility for languages, and melded easily into the college lifestyle at the University of Wisconsin. While there, he earned an MS and became a licensed Wisconsin real estate broker. In toto, Saied ended up staying in Wisconsin longer than I, working and studying in Madison from 1975 through 1987, and enjoying a fruitful career as a chemist and microbiologist.

Finally, toward the end of 1987, Saied was hired by the US Government where he was assigned several different jobs in a number of cities over the

---

* Saied related a rather touching story to me some years later, one of which I had been unaware. Apparently when I had left for America in 1968, he remembers our mother Aida crying herself to sleep for several nights. When he went to comfort her, she confessed it was because I had left. "I will do the same about you when you go," she told him. "And some day you must." Once again our mother had proved to be not only the ultimate empath, but also reaffirmed her prescience, which in hindsight seemed nearly flawless.

coming years—moving first to Alameda, California, then Denver, Colorado, next to Minneapolis, Minnesota, then Washington, DC, and finally to Orlando, Florida. As versatile as he was prolific, Saied held key positions such as a food technologist, food inspector, investigator, compliance officer, supervisor, all the while managing to excel at everything and being promoted in virtually every assignment.

Saied certainly earned himself something of an illustrious career, ended up married and having two children, now in college. He later took a job with the US Government as a food scientist, and became rather well travelled, living first in upstate New York and later in College Station, Texas (noted singularly as the home of Texas A&M University).

Sagacious in his academic pursuits, Saied got two more MBAs at the age of 55 and now works for the department of Homeland Security in Orlando, Florida. Perhaps by intent, though Saied was the one family member who assimilated most readily into the US culture, he is the only one who did not join the rest of the family with me in New Hampshire.

Atif, despite his earlier vicissitudes in the United States, returned at our mother's request to give it another try in 1976. This time, he connected with both New Hampshire and his expanded family there. He tried his hand with his own floor-cleaning business in Lebanon for a few years before he switched careers to work for Dartmouth Hospital in the early 1980s. After 8 years in hospital staff management, my brother moved over to Dartmouth University in their Student Craft/Restaurant Services where he worked for 15 years before retiring on disability.

Mecheal, our youngest brother, was actually the second of my siblings to come to the United States, arriving in Lebanon shortly after I did in early 1976. When he arrived on his own and virtually penniless, I arranged to get him a job as a dishwasher, and later on as a cook at Landers—the very same Lebanese restaurant where I'd had my first date with Bonnie.

Mecheal showed a great facility and understanding for the restaurant business. And after about eighteen months he had risen to become the restaurant manager. Many years later when Landers finally went out of business, he went on to a management position at Village Pizza, one of the most popular Italian restaurants in the area, where he worked for another fifteen years until his retirement.

Both my brothers, Atif and Mecheal have experienced their fair share of health challenges—both being diagnosed with the same Type 2 Diabetes that struck me earlier, and Mecheal suffering additional difficulties with rheumatoid arthritis and physical stress. Atif has had three surgeries on his right knee that have rendered him partially crippled and has had a recent (probable) PTSD diagnosis that has forced him to retire on disability.

My sister Fayza and my mother Aida both tried living in New Hampshire through extended visits for a time back in the 1980s, but neither could relate much to the lifestyle here. The cold weather and lack of ethnic community caused them to long for the warm nights and warm people of Cairo. Like her mother before her, Fayza was more than aware that this was the country where ambitions could be realized and lifestyles could be enriched. So she encouraged both her sons, Nader and Maged to move over here to the United States, specifically to the communities of west-central New Hampshire.

Her younger son Nader took her up on her suggestion and gave it a run over here, but not without some difficulty.

In the beginning, it seemed destined that Nader would come. He had won a scholarship lottery in Egypt. He was attending classes in Hotel Management (similar to those in a University) and was taking to that curriculum with a kind of natural affinity. He had legally managed to bypass the notorious Egyptian military obligation. So that final barrier to his getting out of Egypt had been removed.

Since it was so much easier to get family members over in the late 1980s, we managed to secure Nader's immigration papers to the US where he proved to be a hard worker and a pretty astute businessman. Even though he was cook for a while in a restaurant near Dartmouth when he got here, he was willing to listen when I encouraged him to set aside what small savings he could and use them for investments later.

He even showed more faith in me later when I prevailed upon him to go way out of his comfort zone and buy a small five-unit rental property that I was convinced was the bargain of a lifetime. (To sweeten the deal, I offered a bit of bridge financing to shore up his resolve.) To his credit, Nader bit the bullet (as they say), pumped in every cent he had in savings and prayed that my "vetting system" would prove to be correct.

I'm happy to say the result was fruitful. Nader made back his money on the investment and then some. In a short time, he paid back my "purchase incentive," and still owns the property to this day. He has also been able to apply his hotel management training in the running and maintaining of his properties.

Nader's older brother Maged, though Fayza tried to prevail upon him to move here, remains my only blood nephew who has stayed back in Egypt. Very entrepreneurial and highly innovative as an affiliate with Canon, Maged has developed a way to recycle used business printing materials and resell them at a deep discount—thus providing a boon to the environment and a positive price point for businesses looking to save on their overall material expenditures.

Despite a business misadventure inside Russia when he was younger, Maged has always kept his cultural roots deep inside of Egypt, and still lives in Cairo with his lovely wife and three children.** Frankly, we tend to be concerned for his safety. Given the emergence of ISIS and the resurgence of the Muslim Brotherhood everywhere in the Middle East, Egypt is no longer the best place for Coptic Christians these days.

** Maged learned "the hard way" one of the secrets of the successful entrepreneurs early in his career: "Never get involved in a business you know nothing about." Unfortunately, that's exactly what my nephew did when he decided to move to Russia to open a bakery there and realized two things: 1) he didn't know the market; 2) he didn't know anything about the bakery business. What's more, he didn't realize that Egyptians represented less than .0003% of the population in that country. So he had no one with whom to interrelate. Soon enough, Maged returned to Cairo, a sadder but wiser man. And certainly more realistic about his career choices.

At the moment, Maged lives in Cairo within one block of the Central City Square, the scene of the 2014 riots, a location that sill remains a political tinderbox two years after the fact. And even though Egypt has now returned to some semblance of normalcy due to the leadership of President Abdel Fattah el-Sisi, religious and political repressions continue, and fanaticism is now the way of business in the Middle East.

Of course, we worry about Maged and would like to see him get his family out of harm's way and over to America. By the same token, success is a magnet. He is prospering even in the most difficult of times. And every man's path is his own.

CHAPTER 35

1990s: YEARS OF MOMENTUM

Ask any businessman about the most productive decade of his life. Most of them will give you the same answer: "In my fifties. My most productive years were those in my fifties."

This is a generalization. Certainly musicians, actors and athletes will have dissenting opinions for obvious reasons. Then again, for those of us in the professions, it can also just as easily become a truism. The number, 50, seems to click-on that sense of controlled urgency that might not have been there before. Everything just starts to sync in about that time in your life. Assuming you've attempted to accomplish anything at all, you have usually attained some level of mastery over your chosen profession. You have learned how to use multiple streams of income. And your people skills, depending upon who you are, have become reasonably well developed. By this time in your life you have usually become more of who you really are. So it becomes the perfect convergence of two laws: Attraction and Momentum.

There is the old cliché that holds true: "If you're doing something you love, you never have to work again in your life." My answer to that is "Yes! That's true! Then again, No! It's not!"

Anything worthwhile is actually going to take some work. But there is an enormous difference in working with passion and energy, because it creates something called *eustress*. By definition, eustress is a kind of stress or tension that benefits the person who experiences it. This shows up physically and systemically in people who have won a race or players who've won the Super Bowl or just completed a valuable work of art…or a symphony conductor after the flourish of Beethoven's Ninth. (Eustress actually triggers endorphins and secretes neuropeptides, which in turn build one's immune system that contribute to reconstruction of cellular tissue and eventually go into making you look and feel younger.)

Having said this, I know there is a balancing act. And you can overdo a good thing. So I suppose I'm one of those who never quite got the hang of knowing when to quit while you're ahead.

I do know one thing though. Nothing kills you faster than negative stress. It comes from a thousand sources. And nothing generates more stress than working at something you hate. When you are younger you can work at something you dislike to build toward something you love. When you get into middle-age, if you hate what you're doing you start to grow ill and die. This is just physical law: drudge is a body blow to the soul, and enough blows will pound you down.

By the time I had reached fifty, I was loving my life. I was at full throttle in two careers I enjoyed immensely—my soil science and my newfound bullish involvement in real estate—and I was able to spend some quality time with my family that I cared so much about. What's more I was enjoying the Laws of Attraction if for no other reason than the fact that I was, by sheer force of our body of work, interacting with the finest scientists in my field. And I was now living my philosophy to the fullest—that *Crisis Creates Opportunity.*

Inside CRREL in the late 1980s the fall of the Berlin Wall and the onset collapse of the USSR actually caused us to have to "re-chunk" our focus. Because we were technically a Cold War technology with some peacetime sidebenefits, we actually stood to lose the lion's share of US Government funding when it came to applications from ACE as well as the majority of its military installations. So almost overnight we were experiencing a precipitous drop in funding from generous allocations to practically none at all. At our core, we were facing an avalanche of shutdowns, and cutbacks in staff or transfers would surely be soon to follow.

Please understand that Big Government, by nature, is custodial in its construction. So a government agency seldom if ever becomes entrepreneurial when it comes to creating work for itself. For my part, I was about to change all that when 1989 marked my return to Alaska, this time to "audit" the infamous EXXON *Valdez* incident.

With over 1 million tons of oil spilled into Prince William Sound and the more than $1 billion cleanup that was now taking place, the oil industry in Alaska had become the poster child for economic disaster, and opposition to drilling in Alaska grew to an all-time high. So the Oil Pollution Act (OPA) of 1990 created a set of strict environmental standards and reparations/penalties for energy companies still producing in the area. Included among them were runoff restrictions and reclamation requirements that perfectly fit the frozen ground technology we were working on at CRREL.

Especially with Prudhoe Bay just up the coast acting as the largest producing "oilfield" in North America, we had the primary market in the hemisphere for all our environmental mathematical models.

With being able to apply our frozen ground technologies as a means of freezing, isolating and removing ground pollutants not only for oil fields but also airfields, I saw Alaska as the next major market for CRREL to get both funding and fees from private enterprise, from the military, from the FAA, from the Department of the Interior (the EPA)…from everybody! I was already in an "acquisitive" mindset from my real estate ventures. So I recommended that CRREL open a branch office in Fairbanks to deal with new market demands.

With the military installations in Fort Wainwright, Fort Richardson and Elmendorf Air Force Base all facing either cutbacks or closure, they too were facing large measures of environmental cleanup of poisonous metals from ordinance and military vehicles. CRREL was also being looked-upon by both the FAA and the US Air Force as the single option available to provide "frozen ground" solutions for their challenges. Part of that was a $12 million research fund to deal with these issues. So the notion of a branch office for CRREL became a pretty easy sell.

Beginning in 1990 and for the rest of the decade, Alaska proved to be CRREL's most productive developmental market. And Soil Sciences and Biogeochemistry became arguably the most active and prolific branch inside of CRREL and ACE. And we did it by becoming entrepreneurs in the purest sense of the word.

I use the word, "entrepreneur," advisedly. Better said, we had seen to it that this became a "stable source of income," for CRREL. And our department was leading the way.

Meanwhile, in 1988 and again from 1990 to 1996, our department received Commendation Awards from the Department of the Army for our achievements in Geochemical Sciences, for outstanding performance in establishing new initiatives and environmental quality, and especially for site assessments and reclamation in Alaska. (I say "we" received them. I received the personal awards, but our department—our extraordinary team—earned the kudos for which I was personally recognized.)

In 1990, we set up a Metal Waste Management workshop in Orlando Florida and continued to spearhead the way in frozen ground and wastewater technologies in the Tundra and all the way through the Aleutian Islands off the Coast of Russia.

If I managed to make an important contribution to that time it was to engage our former cold war opponents, the Russians, in active scientific dialogue. Since their permafrost and frozen ground technology was still well ahead of ours in almost every area, it would have been a crime against science and a case of self-sabotage for us not to engage our colleagues there in every possible way. One event that triggered all that was the 1995 International Workshop on Cold Regions Contaminant Hydrology in Anchorage, Alaska, and the meetings that spun out of it.

As Chief of Geochemical Sciences for CRREL, I was honored to receive another performance award for our research and innovation team, our third year in a row. And by then I had long since hired some of the finest minds in Cold Regions Models and Technology such as Dr. Steve Grant and Dr. Giles Merion, and continued to co-author numerous works with Steve, with Magdi Selim, with D.C. Adriano and so many other gifted soil scientists and biogeochemists from all over the world.

We also continued to publish papers on diverse environmental studies. And by 1999, I had co-authored five more published books with Magdi, Steve and others on everything from *Soil Remediation from Contaminated Metals,* to *Artificially Frozen Ground Technology [And Bioremediation].* *

I mention all this not to go on about "my" accomplishments as much as "our" collective body of work. And I'm including a complete chronology at the end of this story if you want to go into specifics.

I am mentioning all this here for a couple of reasons. One had to do with the fact that our explorations into so many areas paralleled what was going on in America—one of the most bullish economies of all time, and an era when everything was possible. The other is to score another point—that the synergy of dedicated people working toward a good result can often accomplish great things.

That is the Law of Attraction at its most complete. And if I can emphasize one talent of mine, it is my ability to locate, recognize, assemble and organize the very best people at what they do and help them to do what they do best. That is at least part of the reason I was able to leverage my position at CRREL to set up and establish so many peer reviews and act as Chairman or Co-Chairman for so many gatherings of the best scientific minds in geosciences.

Some of the landmark conventions we were able to establish included the Fourth International Conference on the Biogeochemistry of Trace Elements at the University of California Berkeley in 1997 (an event that helped inspire

---

* Iskandar, I.K. and H.M. Selim (Editors). *Engineering Aspects of Metal-Waste Management.* Lewis Publishers, Inc., Chelsea, MI. 1992.

Adriano, D. C, I.K. Iskandar and L. Murarka (Editors). *Groundwater Contamination.* Northwood, England, Science and Technology Letters. 1995.

Adriano, D. C., Z. Chen, S. Yang and I. K. Iskandar. (Editors).*Biogeochemistry of Trace Elements, Science Reviews,* Northwood, England. 1997

Iskandar, I. K. and D. C. Adriano (Editors). *Remediation of Soils Contaminated with Metals.* Northwood, England, Science and Technology Letters. 1997.

Selim, H. M. and I. K. Iskandar (Editors). *Fate and Transport of Heavy Metals in the Vadose Zone,* Ann Arbor Press. Ann Arbor, MI 1999.

the establishment and founding of ISTEB). And in 1997, I was honored to Co-Chair the Fourth International Symposium on Physics, Chemistry and Ecology of Seasonally Frozen Soils in Fairbanks. And in the same year, 1997, I was able to help set up and Co-chair, a Fourth International Symposium on Environmental Geotechnology in Boston.

That's a lot a scientific name-dropping, I admit. And yet it underscores the body of work of a great many dedicated scientists. In a way, these events also became a means of expressing my gratitude to my peers and my devotion to the continued global exchange of great new discoveries in our field. After all, this was the 1990s, "The Age of Networking." Environmental science was beginning to peak in everyone's consciousness. And Soil Science, in its many incarnations, was at the forefront.

Well, maybe it was too much of a good thing. (I'll never know for sure.) I can also attest to the fact, here and now, that nothing lasts forever, and that—especially in the bizarre labyrinth of government—people resist too much success. "Creative destruction" unnerves the bureaucrat. And that is what we were doing.

Perhaps not surprisingly after a while, there started to be some pushback inside CRREL from the top management that we might need to reshuffle things a bit. It's never a great idea, if you are a "Branch" of a division of government, to create too much of a stir. The tail starts to wag the dog, and that is a dangerous thing.

That is what started to happen in CRREL in the mid 1990s. In our attempts to take Soil Sciences inside of CRREL and CRREL itself on to new frontiers, we had unwittingly broken old models—very nearly all of them—and I had come to be looked upon as something of an iconoclast; someone who was constantly going against what is known as Conventional Wisdom. (In my experience "conventional wisdom" focuses way too much on convention and not nearly enough on wisdom. So it tended to make me unpopular with certain management inside this *government* organization.)

Before CRREL dismantled its Cold War "prime-directive" in about 1990, it had always been headed up by a Colonel or a Brigadier General from the

US Army Corps of Engineers. And these individuals had always acted as the de-facto CEOs of CRREL.

I always liked military men. They tended to be pragmatic in the best possible ways. If things were working they would leave them alone. If they weren't they would fix them…or find someone who could.

Bureaucrats from the civilian side of government are different animals entirely; they tend to operate from a caretaker mentality. Their primary function is to maintain a kind of gray status quo. So bold entrepreneurial ventures often shake up their paradigms to intolerable levels.

By 1994, CRREL had a new Division Chief who was overtly uncomfortable with the way CRREL was headed and the irresistible force that the Soil Sciences Department, under my supervision, had grown to be inside it.

It seemed almost counterintuitive, because it was in that same year that I personally received a second CRREL Performance Award (for my department). Nevertheless, this new chief and I simply had a difference of philosophy from the word "go." And though he was unable to stop so much of the momentum I had built towards the many aspects of CRREL's newfound versatility when he arrived, I could tell by his gradual micromanagement that he was going to rebuild Cold Regions Research and Engineering Laboratory to fit his new preset paradigm, and a lot of it would involve pulling back on the throttle of what we had built up to now.

Exit paths are strange configurations, especially in large organizations such as state universities and government agencies. They seldom take place for explicit reasons and evident challenges. Instead they take the form of passive-aggressive dismantling, and usually involve transfers of some kind.

My virtual departure from CRREL started in 1997, and it came in a couple of ways. First was the obvious "blocking" tactic that big management applies when trying to force someone's hand. So, suddenly in 1997, I found that initiatives and departmental studies that I had spearheaded before were meeting with resistance, "review," and delays. CRREL started shifting its emphasis away from new exploration studies and more toward holding onto what it had already established. Autonomy granted to me in the past was now being moved upstairs for final approval, often dying there. What's more, I was losing the ability to manage my own personnel.

It is important here to stress that, had this new minion of management even stopped to evaluate our situation for an appropriate time, none of this would have been necessary. Much of the work and new market development we had done in Soil Sciences was a large part of the reason CRREL was still functioning as well as it was. But rather than accept this Biogeoscience Division, this particular individual felt threatened by it, expressing it in ways that could only be termed as passive-aggressive.

For example, I was suddenly no longer authorized to hire my own staff. And once staff was hired, they were denied adequate office space to perform their job effectively. Once in a CRREL division meeting, I made my case to promote a young engineer working under my supervision that I knew had done an outstanding job, only to be summarily dismissed from the meeting. Later my new boss came by to apologize for the insult, possibly because he sensed that I was documenting the insult and lack of professionalism he had shown. Those incidents and others underscored the fact that this was soon becoming an untenable situation for me. The rest of my staff knew this as well, and morale suffered because of it. It was tragic because, over the dozen years I had headed up this Division, we had created a kind of "Starship Enterprise." Now it was being undone.

This came to a head over a young engineer in my division who simply wasn't pulling her weight. She was falsifying her studies, misrepresenting her workload and lying about her actual "hours" devoted to the job. In short, she had become a poison inside the division; my Division.

Encouraged by my own staff to reveal her corruptions, I presented my case and was flatly rebuffed for my efforts. The result was twofold. First, the woman in question was moved over to another division, where a better case of "fraud" was built against her; and she was summarily fired two years later. Second, the true thrust of this event was that I was asked to slide aside as Branch Chief and into a soil science research position.

It was something that I managed to help facilitate because—by that time—I had been invited to be a Distinguished Research Professor at the University of Massachusetts (in Lowell), a level of responsibility that occupied most of my time from autumn of 1997 to the end of 2000 when I finally retired from CRREL…on very good terms.

Of course what has ended never really dies. And this was very much the case at CRREL, simply because my roots in Soil Sciences by now went far too deep. I still maintained a GS-15 rated Professor status. I was still the head of several International Society of Trace Elements and Biogeoscience gatherings. And I was still on the path to being consulted for my expertise in several of these areas (for years).

By the year 2000, however, I was ready to move on, primarily because my entrepreneurial spirit had overshadowed everything else. I had more than paid my dues in academia. For nearly forty years, I had navigated the labyrinths of government in two different countries. And now, I was ready to be out on my own 24-7-365. I was making great money with my real estate investments and stocks. I had discovered that pot of gold called *The 1031 Exchange*. I was well prepared for what we all like to refer to as "The Next Big Thing," not the least of which was being able to spend more quality time with my family.

What that next big thing would be, and the many ways in which it would be expressed, would come as a surprise. The New Millennium was to become an era full of surprises—in ways none of us could ever have imagined.

CHAPTER 36

TRANSITION 2000

There are times when metaphysical law becomes physical. So I'm going to stick my neck out here and say it: Momentum has both mass and energy. I firmly believe that it literally takes on physical form that often dictates the pace of the inertia that surrounds it.

By the year 2000, my life had taken on that kind of irresistible force. To be sure, I had been moved aside at CRREL, thus breaking up the "winningest" team with the most performance commendations in that organization's history. I had also completed my two-year guest professorship at U Mass. That too had been resolved. What hadn't been factored in was the long list of associations and publications in which I had been actively involved.

Certainly, by the end of 1999, I was so immersed in my real estate ownership and property management company that I was ready to make the necessary transition from nearly four decades in Soil Sciences and Biogeochemistry, frozen ground technologies and wastewater management and renovation. The challenge was that I had become so identified with so many areas of that field that when the year 2000 began I was more in demand than ever.

The year 2000 saw the publication of three books related to a decade's worth of research findings and mathematical conversions that I had at least some part in advancing to new levels. If I had gotten really good at some things over the years, this would be my trifecta (so to speak): 1) Getting funding for our research projects (at CRREL or anywhere else). I did that by identifying what financiers call "money looking for project," and by shaping the project to resemble the target that the funding was seeking out; 2) Networking inside the science community—especially geosciences and soil sciences—so that my associates and I had connections all over the world; 3) An ability to get worthwhile projects published—research, studies and bodies of work that might have otherwise gone unrecognized.

While I was in CRREL we had a "publications staff" of some very erudite people who could gather, refine, edit and produce the mass of details, abstracts and reports I was able to bring them. This had been a creative group of gifted researchers and writers. I had seen their benefit and kept them all busy for at least a dozen years. We had established a great working relationship. So, even though I was on my way out the door, I took advantage of this association by placing their services at the disposal of other scientists such as Steve Grant and P. M. Huang whose superb bodies of work needed both advocacy and publication.*

Very often, I was included as a co-author or co-editor for these books when it wasn't something I really sought, simply because my associates insisted upon sharing the credits. My interest was primarily in getting the information out there, so recognition has always been secondary. What I have been something of a bulldog about is actually seeing this valuable information brought to light. So I've never had a problem with rolling up my sleeves and making sure that great research was compiled, edited and anthologized into a proper context. The reasons for this are simple and practical: It is an unfortunate custom in the world of scientific experimentation that the best work by some of our finest minds often gets laid out in a research paper and summarized in a way that only about 20 people understand. Left in that form, it is all-too-often doomed to some level of "peer review" where it eventually dies of neglect. I have always believed that these discoveries—some of them phenomenal—need

---

* Huang, P. M. and I. K. Iskandar (Editors.). *Soils and Groundwater Pollution and Remediation.* Lewis. New York, NY. 2000.

Iskandar, I. K. (Editor.). *Environmental Restoration of Metals Contaminated Soils.* Ann Arbor Press. 2000.

Grant, S. A. and I. K. Iskandar (Eds.). *Models for Cold Regions Contaminant Hydrology: Current Use and Future Needs,* Ann Arbor Press. 2000.

to get the attention they deserve because they have the potential to change lives and more, nations.

At the risk of protesting too much, a great many of these books might never have made it into print had someone not undertaken to guarantee their completion by offering a certain degree of both elbow grease and subsidy. I suppose I had to be that "someone."

For me there was just too much at stake, and I didn't want to leave the jobs unfinished. In the case of P.M. Huang, I had begun the journey with him years before through mutual associations in Soil Sciences through our doctorate degrees at the University of Wisconsin in Madison. We knew each other only casually but had shared goals in many areas that had kept us in touch over the years. I had been impressed with his brilliant work in identifying the heavy metals in ground pollution and the fact that was he able to determine through mathematical models that a great deal of heavy metals in groundwater could be dissipated simply by changing the kind of soil in which certain crops were grown.

Big Agra knows (and the general public suspects) that our pesticide/chemical fertilizer/machine extraction system of factory farming has for decades been contaminating our food chain, so much so that the food we get on our tables contains trace-elements that slip in just below safe limits. That means they are marginal and therefore dangerous at every level of consumption. Dr. Huang had shown, through studies in Australia, that the soil itself—provided the groundwater traveled far enough through it—could be an adequate filter to re-create a healthy growth medium. (He just couldn't get anybody interested enough to publish it.)

I took this on as a cause célèbre and added some of our findings into the mix, took it to my publishers at Ann Arbor Press and guaranteed a certain number of copies, knowing full well that if these studies reached enough of the right people they could improve our understanding of healthy growth media a thousand fold. In certain circles, we were able to reach the right people, and it ended up making much of the impact I'd hoped for.

A similar situation applied to the research done by my young friend and associate at CRREL, Steve Grant. I had mentioned Steve's work earlier in this book, but needed to emphasize the point that he had become an expert in what is known as *contaminant hydrology and permafrost solutions* for isolating toxic wastes, especially what biogeochemists refer to as LRW *(Liquid Radioactive Waste)*.

A dedicated environmentalist, Steve had put together an impressive set of findings and come to some outstanding conclusions, most of which were put forth into a couple of studies that would have been read by a half-dozen people at CRREL (especially in the new regime) and probably left to die there. Since his findings very much complemented some of the studies I had originally espoused and helped expand upon, I felt very strongly that this superb body of work needed to be taken seriously. So, once again, I put on my publishing hat, helped Steve pull all this together and guaranteed a certain level of sales to our friends at Ann Arbor Press. The result was a focused examination by some of the most strategic environmental scientists in the world…and a major topic of study at what was to be our next ISTEB Convention in Uppsala, Sweden.

Understand, and I emphasize again, that the cold regions modeling technologies we were proposing for study and possible adoption had the ability to provide three-step solutions—*isolation, detraction and removal*—for most Liquid Radioactive Waste challenges anywhere in the world. Had they been in place as preventive "frozen ground" barriers at Three Mile Island in 1975, in Chernobyl in 1986 and (especially) in Fukushima in 2013 we could have avoided three of the worst nuclear reactor disasters in the last 100 years—the last of which may have polluted the entire Pacific Ocean, certainly to within an 800-mile radius of ground zero.

Now, of course, these measures are in place in Fukushima…after the fact. Had it been in place and available at the turn of a tap, I am convinced this disaster could have been prevented or at least cut short at a much earlier level. And I, for one, believe it should be mandatory wherever LRW, massive oil production or military ordinance are constantly present. All this work and more had carried over for me well into the year 2000 when I finally stepped aside and officially retired from CRREL.

There is little question in my mind that at least part of my mutually agreed upon departure at CRREL had a great deal to do with our initiatives and discoveries on groundwater remediation, heavy metals conversion and frozen ground technologies—not because we were conducting them but because we were directing them toward *true environmental solutions*. To be perfectly frank, unless the light of public scrutiny shines on it, 95% of the time environmental reclamation is never a part of government focus. As far as our original function and its relationship with ACE were concerned, CRREL had always been primarily a research lab. And, to say the least, my environmental activism had shaken up the status quo.

Ironically, The New Millennium also marked the same year I was voted in as President for the International Society of Trace Elements and Biogeochemistry (ISTEB). So while I was leaving the microcosm of CRREL, the macrocosm of soil sciences and trace elements had now extended itself to me on a truly international level.

This new incarnation for me enabled me to function more extensively in the two areas I had truly come to enjoy more than any: 1) actively networking new discoveries in Soil Science and biogeochemistry; 2) my new role as CEO and General Manager of Lebanon Property Management.

By the end of the year 2000, I felt I had paid my dues in the worlds of government and academia. Although I would miss my associations there, I wouldn't miss the intrigues, the internecine politics and the intellectual piracy that often took place in these minor galaxies of endeavor.

Now that I was on my own, I reasoned, I would be the master of my fate. And I wouldn't have to deal with so many of the turf wars and self-justification that seemed to prevail in these worlds. Of course, I couldn't have been more wrong about that. But one thing I'd gotten right: I could take responsibility for anything that happened from that day forward. And once you are prepared to embrace that ability, you can create your own realities.

My new realities included the joy of ownership. And by the beginning of 2000 I had used my profits from the 1031 Exchange to make a series of new acquisitions.

The year 2000 was a turning point for many reasons. For everyone, it represented the boom years of the Clinton Administration (helped by a Republican Congress), a technology bubble in the stock market and a relative sense of invulnerability and the rise of the Internet as the Alpha and Omega of all commerce.

This all came crashing down on September 11, 2001 along with the Twin Towers in New York.

I remember seeing the nightmare that unfolded before me (before us all) as it was broadcast and rebroadcast for two days, showing how the towers at One World Trade Center came crashing to the ground. I remember too how the speculation came over the major networks about who the perpetrators had been, when I personally never doubted for a minute that the sabotage had come at the hands of Radical Jihad.

It was the gradual Islamization of Egypt that had actually driven me from my own country when I was a young man. Subtly begun during the Nasser regime, it intensified during the decade of the Sadat administration to a point of institutional discrimination at every level. I was made especially aware of this during the trips I took back to Egypt, Saudi Arabia and nations to which I had travelled in the 1990s to lecture and visit family.

So, even though this came as no surprise, I couldn't help but be repulsed by the magnitude of the atrocities of 9/11, just as I was heartened by the show of patriotism and unity Americans displayed in the days that followed. (It certainly solidified mine.)

I had mentioned the irony of my encounter with the FBI in the wake of this event and the days that followed the disaster at the Twin Towers. And I was only too happy to lend my gifts of translation to help our intelligence agencies with any translations from Arabic they needed to interpret in the coming years.

Not surprisingly, 9/11/01 also created a paradigm shift in the American business psyche in the coming years. We went into what amounted to a wartime economy virtually for the next decade. The Stock Market tanked for nearly a year. (This time, I held steady and kept the investments that I knew would hold.) And there was also a kind of "false spring" for real estate that came to accelerate around 2002 with FNMA ("Fannie Mae") *teaser rates, interest only, negative amortization loans* and easy money for people who wanted to buy homes.

This had officially started during the Clinton Administration with the Community Reinvestment Act of 1999—something rammed through by a "Bull-Market Congress" when everybody seemed liquid. Later, after 2001 when the stock market hovered around 9000 for months and the Bush Administration was trying to jumpstart the economy, easy homeowner loans were becoming the order of the day, and Fannie Mae and Freddie Mac money was flowing like water.

I tended to stay away from those easy money deals because I thought they were fraught with pitfalls. I had remembered how inventive I had to be just to get my first properties. And I had seen enough mistakes made over the years to know that this might very well be a bubble that was going to pop somewhere along the way. Still, I also realized that the areas around Hanover and Lebanon were underdeveloped.

Around 2002, I had just sold my two properties in Florida as well as business center in Lebanon for a healthy profit, and I needed to turn them around on a 1031 Exchange within the allotted time to avoid excessive taxes. So, I was looking for an acquisition with a great deal of upside potential, what people in the world of acquisition call a "sleeper." Well, I was looking for a sleeper, and a *Sleeper* is what I got—in so many ways I never imagined.

CHAPTER 37

'SLEEPER'

The French philosopher Voltaire once wryly observed: "If you want to make God laugh, tell him your plans." Nothing validates that cynical point of view more than my wonderful, terrible, exciting, madding experience with a little project that I ended up ironically naming "Sleeper Village," after the last farmer who had farmed that land.

Actually, this purchase was going to be my happy place, my retirement into the gracious life of the gentleman farmer. In fact, it became anything but that.

As I have mentioned before, I credit the US Government and the IRS for helping to make many of us millionaires simply by challenging us to learn all the ins and outs of the IRS tax codes. Using a large portion of the monies I had made from selling my Stone Farm Apartments on a 1031 Exchange (around $960 K), I purchased a plot of land lying inbetween Lebanon and West Lebanon.*

It was 330 acres of farm and forest primeval formerly owned by a wonderful farmer and local hero, Marvin Sleeper, that I had set aside for development in some different and very creative ways to section off into three expertly designed, superbly planned, dynamic, modern and very livable communities.**

---

* I had actually bought the 330-acre tract of land at the end of 1999, but didn't set the plans in place to develop it in its entirety until 2002.

** Of course there was also to be a lovely 10-acre plot of land with a farmhouse and a large red barn sectioned off for me to retire, grow a victory garden and feed my chickens and my goats to my heart's content.

At this point, I emphasize the fact that this was not just some super-developer scheme. It was an idea whose time had come—not only come but also one that was long overdue. In truth, by 2002, the boroughs of Lebanon and Hanover New Hampshire had become a prohibitively expensive seller's market. Young single professionals and young families in particular were getting squeezed into high rent properties that offered little value for money. They were constantly scrambling to find housing in this area and were either being turned away for lack of available space or were being priced out of any reasonable range to buy, lease or rent. So they were just moving on to other areas. The community was losing income and growth potential. And there seemed to be no end to it. So a new development was truly needed to inject some energy into the area and solve some serious housing issues that were simply not being addressed.

At the outset, I had planned to call the development Sleeper Village in tribute to that very fine gentleman farmer, Marvin Sleeper. And it was part of my vision for what seemed to me to be some of the most underdeveloped areas in the entirety of New England. The 330-acre area was to act as home to three distinct communities broken off into 450 exquisitely planned units:

- 150 homes and duplexes on 31 acres ranging from three-fourths of an acre to two-and-a-half acres.
- a 150-unit apartment complex catering to singles and young couples.
- a 150-unit retirement village for seniors, including town homes and cottages.

Not just thrown together, these were to be model communities replete with parks, walkways, shops, waterways and a church (worship center)—all making optimum use of the natural forest, rivers, streams, soils and surrounding terrain in the area. I had made a feasibility study and had engineers and architects plot the utility factor of the properties and even design mini-model units and the grounds that were to contain them. And I had it put into a very elaborate "futuristic" presentation.

So, one would assume the city fathers and landowners would leap at the chance to bring this kind of dynamic, prosperous futuristic community to the area around Lebanon and Hanover. One would assume…

There is an ancient code of conduct called The Four Agreements that I recommend to anyone as guideline for the way they negotiate their world. The First One is, "Be impeccable to your word." Number two is, "Never make assumptions."

Well, while I was off trying to follow the tenets of the First Agreement, I had not paid the necessary attention to Number Two. I had taken all the right steps, I had thought, to get the project under way. I had gone to the Lebanon City Counsel, worked with the city planners, and gotten permits to construct the houses. What I had overlooked were the new zoning laws between Lebanon and West Lebanon that I discovered to my dismay would have to be rewritten in their entirety to allow me to get this mega-development underway. That would take some serious "missionary work" with the local property owners to convince them of this bold new initiative whose economic and environmental benefits should be evident to anyone with eyes to see.

As I noted before: "Never make assumptions." In this case, my initial presentation to the neighbors and the property owners associations fell absolutely flat. They just didn't get it. Not only did they not get it, they saw this as some insidious get rich scheme on my part that came with some (yet unseen) sinister implications.

Apparently, even though I had been a citizen in good-standing this community for more than 25 years, I was still looked upon by some (not all) as this money-grubbing outsider trying to amass a fortune at their expense! They didn't understand; at least not yet. But all was not lost as far as I was concerned. There was still a chance. With new elections staring us in the face, we could present a bond issue in the next city election in 2002 that could appropriately advance our cause, one that would come to be known as Proposition 20.

Before I go on, I might mention the other two of the Four Agreements because they help to complete the Code from which I was trying to operate: Agreement Three: "Never take anything personally." And finally Number Four: "Just do your best."

Even though my friends, my daughter Amy (who was now locked into a business career with our management company) and even my wife Bonnie thought the homeowners' rejection of our plan to be very personal, I chose to think otherwise. Against some evidence to the contrary, I remained convinced, if I could just state my case clearly and with the right graphics that reason would prevail and the citizens of Lebanon would come around.

I knew this much: It was merely fear-based behavior. People have an almost irrational fear of change, until they are shown the light. And I believed, if I could just put my best foot forward, I could bring the naysayers over to our side.

What followed was certainly our best effort. With Amy's help and with the advice and counsel of former city engineer Dan Nash and others, we designed, assembled and distributed brochures, put on power point presentations and launched an advertising campaign showing the local townspeople all the potential benefits of "Sleeper Village." It would not only make optimum use of the land and surrounding environment, it would improve it. It would bring 500 new jobs to the area and add income to the local economy to the tune of at least $40 million in the first five years.

What's more, it even offered a solution to the traffic jams that were loading up on Route 120 and backing down through into that area from the Exit traffic on I 89. (In a seemingly quiet community focused on both access and privacy, this had become a nightmare, one we felt was unnecessary. And we had come up with a solution to fix it that I was willing to pull funds from my own pocket to resolve!) To my way of thinking, with our Sleeper Village PR campaign we had provided a flawless feasibility study that any fifth-grader could see was right for Lebanon. We had outlined every last detail of the benefits this development would bring. And it lost by about 121 votes.

After Proposition 20 went down in flames and our dreams of Sleeper Village with it, several people came to me later to apologize for their opposition to the initiative. Many of them revealed the fact that they had been influenced by a few people who were convinced of my larcenous intentions, and my unmitigated greed. I was guilty of no such thing of course. My greatest transgression had been to mistake the character of a community by naïvely presuming that everyone wanted the same bold and beautiful future for Lebanon, New Hampshire that I did.

New England communities, I have since come to learn, are almost protectionist in their tidy perceptions of what constitutes progress and how it should be handled. So any change at all comes as a threat to their parochial sense of community.***

Later, in 2005, in the ultimate irony I sold the land of Sleeper Village (with all permits) to an entrepreneur/land speculator from out of the area for more than five times what I had paid for it. He changed the name from Sleeper Village to Rock Ridge Estates, ultimately defaulted on the loan he'd secured to buy the property and ended up going into foreclosure and auction, a comedy of errors that ultimately cost the taxpayers nearly $11 million in the bargain.

Out of what seemed a predatory urge to zero-lot-line the world, this latter-day developer's first initiatives for Rock Ridge Estates were going to be doubling-down on the gracious, spacious community I had set forth in Sleeper Village and cut all the lots into one half to one third the size. In doing so, he would have built up a jungle of jammed together houses and

---

*** Of course there were seemingly dozens of objections to our initiatives in Prop 20; none of them really valid. In the beginning, when hit with the proposal, they worried that their property values would drop; quite the contrary, they would have doubled or even tripled in some instances. They feared that the area would lose its "scenic route" status and that the thruways would create traffic problems; in truth they would have improved them. What the experience did teach me was that I had to get the landowners and neighbors involved from the beginning. That is what I did later with a new development called EvenChance. And that is the rest of the story…

created the very nightmare that the locals had been unrealistically fearing from me. Fortunately, he never got the chance to do so.

The bottom line to all of this was certainly a lesson well learned. I had lost my opportunity to retire as a gentleman farmer (and frankly lost any real interest in retiring at all). I had also been able to apply my philosophy that Crisis Creates Opportunity, and had remembered The Third of the Four Agreements above all else: "Never take anything personally." I didn't. And I did…move on to something even better.

Two distinct side-benefits sprang out of the Sleeper Village saga for which I am extremely grateful. The main one for me happened to be the emergence of my daughter Amy as an excellent, astute business mind, a dedicated associate and loyal advocate.

Having earlier spent a couple of her "college years" travelling the country and soul searching, Amy returned to Lebanon and revisited her childhood inclination to embrace the machinations of business, especially real-estate and property management. She proved to be a bulldog and a determined partner during the entire Sleeper Village experience and even took "an exploratory job" with the new group that bought Sleeper Village while they were trying to turn it into "Rock Ridge Estates." (They were really quite anxious to hire my daughter, and I thought it was a good idea—to help Amy get some exposure under her belt.)

After a few months of experiencing what she described as "learning all the ways not to do business," Amy returned to our group to become the youngest person in the State of New Hampshire ever to get a real-estate broker's license. For the last several years, she has managed the daily operations of Lebanon Property Management to great effect. Other than my organic partnership with my wife Bonnie, Amy is the only other family member who has chosen to go into business with me, proving to be an invaluable working associate along the way. And she shows great instincts for doing what I call "smart business with integrity."

The other side benefit started out as a challenge, but like so many challenges turned out to be the ultimate learning experience and a great guideline for future business. That, for me, came by way of putting in place the final piece in the unfinished puzzle of the 1031 Exchange. That had to do with the identity of my next like-kind "entity" on the heel of this highly profitable sale.

Anyone who has ever done business on a large scale—particularly when it comes to buying and selling property—knows what a minefield the tax codes can be. You not only have the IRS to deal with, you also have to face tax codes and caveats at the state and local level, some of which are even more difficult. That is why we set up things like LLCs (usually for a single project or production), "S" Corps and "C" corps.

That hit home in spades with the sale of Sleeper Village in 2006, when Bonnie and I received notice that the "like-kind" 1031 transfer of our profits from Sleeper Village to the purchase of the 45 Lyme Road (Office Complex) would not qualify for the "exchange program," and that we would in fact have to cough up $650,000 in capital gains tax for the state of New Hampshire.

"What's in a name?" Shakespeare once asked. We learned the hard way that, according to the state of New Hampshire, it meant a great deal. Their interpretation of the "like-kind" sale of a property was much more literal than the IRS and much less forgiving. In fact the same company or financial group or individuals—according to the New Hampshire tax code—had to be *exactly the same "tax entity" by name*. There could be no transfers of identity even if it involved the same group, corporation, partnership or individuals.

Unwittingly, with the approval of our lawyer, Bonnie and I had converted our "like-kind" entity from a family trust to a new LLC formed just for the 45 Lyme Road property, and the New Hampshire tax codes didn't like it one bit. (Please understand, this had already been approved by the IRS. So, to put it mildly, the bill from the New Hampshire Department of Revenue Administration for $650K hit us like a broadside.)

Without risking the obvious and encouraging anyone and everyone to "read the fine print," I will emphasize this one thing: "Always be ready to be your own lawyer, and your own CPA."

Qualified people can only help you so far. Ultimately, it's not their money; it's yours. So treat it accordingly, as if your life depended on it; someday it may. Also be your own counselor. I was in this case, and it proved to be the right decision.

Doubtless, in the case of this transaction, some mistakes had been made. Bonnie and I had been advised incorrectly. We stood to lose several hundred thousand dollars, and fingers were pointing in every direction to assess blame. I was even encouraged to "sue some people." (Not my style at all.)

Litigation is a loser's game. The only people who make money in a civil suit are the lawyers. It creates bad energy, ruins relationships, and wastes a lot of time you could be spending more constructively on something else. So, I decided to use both leverage and patience…and to do a little politicking.

Fortunately for me, New Hampshire is a very small state. (It's not New York or Texas or [God forbid] California.) If you've made any positive impact at all in your community, you know someone at the state level who might be able to help you. So I decided to use that leverage. (Part of that meant going back to our original §1031 Exchange QI George Foss, and teaming up with him to do a little liaison work with the appropriate parties. George had proved himself to be more than trustworthy over our long association, and he also new some of the ins and outs of New Hampshire politics.)

Realizing that the Tax Codes of New Hampshire Revenue Administration were counterintuitive to those at the national level, I invoked the US Internal Revenue Tax Codes to prevail upon our state legislature to review and hopefully renew their paradigms. (After all, they just wanted the money, and "some was better than none.") By working behind the scenes and making some friends along the way, I managed to bring this blind-spot in the New Hampshire Tax Code to light, and even managed to get the blessing of New Hampshire's new Governor John Lynch. †

---

† I actually got my photo taken with "Jack" Lynch, who was recently voted the most popular Governor in New Hampshire history.

The happy ending to all this is that the New Hampshire legislature voted to modify its tax codes to match the parameters set by the Internal Revenue Service of the United States. We were allowed to use our new LLC for the Lyme Road business complex to make the 1031 Exchange, to do so with a minimum tax contribution, and without future recourse. So, in future, anyone who was to make these exchanges for their investment properties would benefit from the rewritten tax codes to favor the transferees.

(You're welcome ☺)

CHAPTER 38

45 LYME ROAD: THE REVERSE EXCHANGE

There is an addendum to this story that proves, more than anything that real estate tax transactions are not for sissies. As if the Sleeper Village Saga weren't complicated enough, I had an interesting challenge with 1031 Exchange on 45 Lyme Road that gave me an education I hadn't anticipated but have since learned to appreciate.

This took place because of what I have to refer to as "The Nervous Seller Syndrome." It is pure *Fear Matrix* behavior that strikes people in real estate transactions all too often, but one I've learned to accept as part of list of potential pitfalls that might occur.

When we went to make the traditional 1031 Exchange transaction—selling Sleeper Village and using the transfer of funds to purchase 45 Lyme Road—we did so with the anticipation that all parties would agree to the normal timeframe required for the exchange to take place (within window of 180 days). Unfortunately, the seller of 45 Lyme Road showed absolutely no inclination to accommodate our "tax timetables" and demanded the money within 45 days or he would pull the deal off the table.

This left me on the horns of a dilemma, because if I failed to meet his cash demands, the transaction would go sideways. I would have to leap outside the careful boundaries set by the "like-kind" transaction required under the tenets of 1031 and would lose on both ends of the contract.

This challenge to my cash flow and completion of the exchange was further complicated due to the fact that, at the same time we were completing our Lyme Road transaction, I was also closing on the sale of a 2.2 acre property on Route 120, half of which was to be owned by the local Credit Union, and the other half (1.1 acres) for which I was to construct and rent what would soon be known as the Morgan Stanley Building. This enabled both of us to jointly own the land, allowing the Credit Union to enjoy ownership of their lot and building, which was as much a good community gesture on our part as a solid business deal.

After lowering my purchase price to the Credit Union to conform to their budget restrictions, I was due to complete the sale, from which the funds were to go in turn toward the final purchase of Lyme Road. But the "rush job" I was getting from the nervous investor at Lyme Road was causing problems at all three locations. Add to that the fact that we had subdivided the property at Route 120 to 1.1 acres (the other half of the land in question being designated to build a new building for Morgan Stanley). Since this kind of joint property holding was out of the ordinary, the subdivision would require ratification from the Lebanon Planning Board. (And as anyone who has ever dealt with planning boards anywhere can tell you, their approval pace is glacial…because they can!)

So here we had the worst possible transactional trifecta: a panicky investor at Lyme screaming for his money, an obstructive Lebanon Planning Board taking their own sweet time, and a potential tenant at Morgan Stanley whose lawyer wouldn't commit…to anything! With due dates for all properties closing fast, if indeed we were to qualify for the 1031 Exchange tax credits. For me it was the cash flow Perfect Storm, and I stood to lose a bundle… until I learned the ramifications of what we now call the Reverse [1031] Exchange.

I won't go into the specific amounts, but the implications on all three ends amounted to nearly a $3.5 million swing in a direction not exactly in our favor. So, we had to act quickly, and do so with someone who actually knew how to unravel all the complicated wrappings of what had become a "Chinese Box" of timetables, cash flow, egos and possible zoning hiccups.

One of the philosophies I have embraced over the years is to trust smart people once they have earned my trust, and to go to them for advice. (This applied to my experiences in geosciences. And it certainly applies now to all my dealings in real estate and finance.) Knowing this Lyme Road/Sleeper Village exchange might present a real financial minefield; I went back to my original Qualified Intermediary George Foss with my dilemma. And after examining all the options on the table, George responded in a way that surprised me.

"No problem," George told me. "We'll work a §1031 'Reverse' Exchange."

Apparently there is a reason business tax codes are so elaborate. Some very smart people factor-in everything and do manage to cover every eventuality.

In this case, many reasons have cropped up over the decades for setting up a Reverse Exchange, but the major consideration is the critical issue of finding a way to take ownership of the replacement property prior to the sale of the original property in the 1031 exchange.

Since the IRS tax code originally does not allow for the exchanger to exchange into a property already owned, the Reverse 1031 Exchange becomes the best answer when the seller is ready to close on the replacement property while still trying to sell the old, original property as a means of securing your replacement property. You do that to reduce the risk of possibly losing that property. Also, this way you rid yourself of the replacement property "dilemma" once you have sold that old (original) property because there is a short 45-day window to find a suitable replacement investment on a relinquished property, unless the purchase of the new property is made.

There were about three different ways we could go about this reverse exchange, and the one we chose was something called *The Safe Harbor Reverse Exchange*.

This is a transaction whereby the Qualified Intermediary takes control or "parks" the replacement property in an LLC [Limited Liability Company] or LP [Limited Partnership] under his or her aegis prior to the sale of the old, original property. The exchanger is forced to identify the relinquished property within the 45 days of the parking arrangement. And they must also have all transactions completed within 180 days of the IRS approved parking arrangement.

In this case the Safe Harbor Exchange entailed a double-transaction, also allowed as long as the 1031 Exchange is made on a property or properties of the same or greater value than the property being sold. And by setting up an LLC— Acorn LLC in this case—we were able to use this entity as a "parking mechanism" until the timetables for the exchange were officially met. With the establishment of Acorn LLC we were allowed an intermediary transaction whereby Acorn was able to purchase the (proposed) Morgan Stanley Building at Route 120 and bundle it with the proceeds from Sleeper Village. So, in fact, a series of dummy corporations (Acorn 1, 2 and 3) were set up to hold the money transfers—as "like/like/like" transactions—until the transfers were complete.

This structure falls within a revenue procedure set forth by the IRS tax code in the year 2000 and it also provides the guidelines in structuring this type of transaction or exchange. If it is properly structured within the provisions of the revenue procedure, the IRS will treat this as fitting within a "safe-harbor." They will not challenge the transaction based upon its status as a Reverse 1031 exchange. This is the most secure type of reverse 1031 exchange, and yet it is also the most difficult to accomplish due to the short time frames that we were working with.

What it required from me to pull this off was plain and simple: I needed to plop down the cash for the entire purchase of 45 Lyme Road prior to the original completion of the 1031 Exchange from Sleeper Village. Mind you, if someone was underfunded or couldn't get the financing, they would have lost on both ends of the transaction.

It also necessitated my fronting the funds for the Morgan Stanley negotiation on good faith that they would honor their (verbal) agreement…and set up a construction platform to complete the structure before the onslaught of winter and building-crew costs that would surely shoot through the roof. So pulling this off required nerves of steel and an extension of good faith from all concerned—most of all from me.

Without going into all the steps made through the reverse exchange, I can only revert to that ultimate cliché: "It's complicated." Ultimately by allowing Acorn to work through the Planning Board to get the building approved and by applying the mortgage payments to be made by means of a series of buyer and seller "Stamps" we were able to pay our minimal tax fees to the State of New Hampshire and get the transactions completed on the established due date. And yet all this was allowed to happen only because the perfectly engineered structure of the Reverse Exchange was set in place in the appropriate time frame. Trust me, this is advanced Real Estate Law 404, and is not for everyone—certainly not for the faint of heart! It took some true grit, solid financial histories, and a dash of good community relations to pull it off.

In the end, our salvation throughout all this elaborate "triple play" was my own frugality. Finally it paid off! I had the available funds to match the purchase price. My QI, George Foss, knew I was liquid. So we could satisfy this rather demanding seller at Lyme Road by making an expedient cash purchase, and still keep the integrity of our "1031 Exchange Qualification" with the Internal Revenue Service.

The bottom line is that everyone went home happy. The purchase and construction of the Morgan Stanley Building had been successfully consummated. The seller at 45 Lyme Road completed his sale in a manner that suited his sense of "urgency." I was able to meet all the criteria the § 1031 Exchange on 45 Lyme Road. And Lebanon Property Management still has a Blue Chip office complex under its umbrella to this day—three of them in fact.

CHAPTER 39

EVENCHANCE: A PRACTICAL VISION

I have always trusted in planning for the future, and I try to apply it in everything I do. Whether it is my personal development, my scientific research or the community to which I am wholeheartedly committed, I believe in setting paradigms that are 100 years or even 200 years ahead. My daughter Amy chastises me for that from time to time and tells me I'm venturing a little too far out there into the unseen mystery of coming events. So I'm encouraged to rein it back in to a more realistic timetable—like 50 years.

The important issue (and the keyword here) is something like having a vision for the best of all possible worlds. Once a real estate broker, developer or property owner is going to work in a town or city for what amounts to decades, he or she would do well to have a sense of destiny where her or his immediate community is concerned.

Individual units—homes, offices or condos—have their own integrity and their own needs. On a larger scale, the purchase and sale becomes organic and almost cries out for proper planning and an appropriate sense of the future. Rivers, ponds, lakes, roads and streets, traffic patterns, access lot size, space, scope and privacy—all these aspects and more have to be taken into consideration if the area is to thrive at all. And whatever group or individuals working on this need to have a sense of what will work decades down the road.

I've always had an ingrained sense of obligation to creating that kind of social structural framework. I feel that any place to which I am going to commit my energies should epitomize both quality and flow. It needs to be the kind of place that people who live there will know was designed and created with everyone's best interest at heart.

Sometimes that comes with great difficulty, especially if you don't understand the process. And part of the process is to understand the degree of change an area will tolerate; not so much in fact as it is the perception of fact. The primary difference between shock and progress becomes one of planning, education and involvement.

Experience can be cruel teacher. What it taught me when we set forth with our grand design for Sleeper Village was not to go too far out. Upon reflection, I engaged in something called "Empire Building." (No, this is not the kind of *empire building* where one's plans and investments outstrip his ability to meet costs. There is that other kind of *empire building,* the kind that takes people too far out of their comfort zone.) Although I firmly believe that the community designs for Sleeper Village were a master plan for the future, I have learned some very good lessons along the way.

One cardinal rule is to *Know Your Market.* You can build a development in Dallas, Texas or Denver, Colorado or Portland, Oregon that involve hundreds of units housing thousands of people, and no one will think a single thing about it. It is part of the dynamic of "life in the big city." People expect it and accept it.

If my Sleeper Village saga taught me anything it is that people in smaller areas, small towns and rural America, are living in those locales for very different reasons, not the least of which is their right to privacy and a calm, unchallenged bucolic lifestyle. So, there are times when you simply cannot superimpose a really grand plan on smaller markets—at least not without going through several very careful steps.

The first one is to work from the inside out. Since the future of any community involves all the members of that community, the first people you work with are the ones who are going to be directly affected by your vision—friends, neighbors and property owners immediately involved in the process. The second is that all great planning is a matter of evolution not revolution, of cooperation and not coercion. Plan, develop and create all within a matter of careful steps and involve everyone else along the way. The third follows what I have always believed: Do your homework: learn all the ins-and-outs of what has to be done. Fourth, and perhaps most important, is always to plan and develop properties that people can actually afford.

When I got the idea for a small community we call Evenchance, I was able to take all the lessons learned from Sleeper Village and reform them into a game plan with a properly developed environment like the one I have just described.

Evenchance got its name from the fact that most of the properties in question were situated on or around Evenchance Road along the southwest coast of Lake Mascoma, near Lebanon. Lake Mascoma is a natural lake and one of the largest in New Hampshire, flanked by the towns of Hanover on the North and East and on Enfield and Lebanon on the West and Southwest. The lake itself plays host to the Dartmouth University Sailing Club and a community yachting club called Shaker Village.

Real estate surrounding bodies of water is always desirable, especially when it is a beautiful lake that hosts sailing and feeds into rivers such as the Mascoma…and ultimately the Connecticut.

In this case, we were able to overcome some zoning restrictions as they related to sewage lines by applying some of the cold regions "frozen ground" technology we had learned to master at CRREL. In this case, it was incorporating my original design for a high point-of-origin, directional drilling technique that would transfer all sewage away from Sleeper Village and around Mascoma River without directly affecting the adjacent land or bodies of water. In the case of Evenchance, I was able to discourage the original developer by convincing him to remove from consideration any individual wells or septic tanks that would inevitably seep into the surrounding terrain and possibly pollute any innate potable groundwater. To eliminate that obstacle to our progress, I purchased an easement from a neighbor that would allow us to connect directly with the city sewer pipes of Lebanon.

Originally, Evenchance was 180 acres of land sloping toward the lake that had been purchased by a friend of mine. I, in turn, bought the same 180 acres from him, setting aside 30 acres that are closer to the lake itself where we have been planning to place 17 homes on lots that average 1½ acres each, leaving one home under development as a landing point for the remaining 150 acres, which may, over time, be used for *other purposes.**

Without going into all the details and specific costs of purchase and resale, I can safely say that this is a wonderfully futuristic development in which all of the neighbors, prospective owners and I have been partners in both the progress and the projections. From the beginning, I got input from the neighbors before I ever submitted the proposed development to the planning board of the township of Enfield. In so doing, we have answered every potential challenge, including sewer lines, access roads, and private streets.

One of the side-benefits of my decades long involvement in soil science and biogeochemistry has been my ability to evaluate the native terrain and foresee any possible challenges. I had been able to use it in the past to evaluate the difficult (limestone) terrain around my earlier purchases at Stone Farm and Mascoma River. And it served me well here.

We were also able to maximize the appeal of the property by allowing the streets and roads leading into and from Evenchance to remain semi-private, and not be turned into a thoroughfare as such. By doing this, we have helped enhance the appeal of the property, have respected the privacy of the residents, and have maintained an integrity of balance between modern progress and Mother Nature. I have even gone so far as to buy 400 feet of waterfront property, including a boat dock, to allow all the neighbors of Evenchance free equal access to the lake—its boating, sailing and fishing.

Finally, we have made the purchase price of these one-to-two acre lots affordable by selling them for no more than $100,000 apiece. That way, we

---

* One of the other purposes for which we have consigned the additional 150 acres is to the township of Enfield that is now drilling through the soil media for a "pure water source." If it is tapped into, that pure water well might be used as a production site for a natural water well to be developed for commercial drinking "designer water," for which we would receive a small royalty.

enable our buyers to determine the level of their dreams, and the kinds of homes they would like to put on these beautiful natural grounds. Even given my very modest purchasing price—by the time we make the road, sewer and zoning adjustments that we are funding out of pocket, and after every lot is sold and developments undertaken—our profit on the entire development will be proportionately modest.

In a way that comes with a degree of intention. Since I have never been comfortable with construction as such, especially dealing with contractors and subcontractors, I won't have to be dropping myself into an aspect of property development that has never really attracted me. Perhaps even more important, by investing my own funds into maximizing the "livability" of this new development, I am able to give something back—to the community in which I have flourished, to the town where my family has come to enjoy all that is best about America, and to the gates to new possibilities…for all our dreams to come.

The difference in what we're doing now—through Evenchance—is that we have learned all the right things to do so we could avoid all the pitfalls of what doesn't work. The difference between what is taking place now and where we missed the boat with Sleeper Village is that we know all the players with whom we are dealing. Another way of saying it might be to use the sports analogy: because this time we have "played inside the lines." Since 2002 in particular, I have become an acknowledged member of the community, recognized by city leaders, the Planning Board and my neighbors alike as someone who shares a common interest with them, and a common purpose.

At this point in my life, I have been blessed and fortunate to have made a pretty penny in real estate, at least to the degree that I can recognize what I do best and embrace the fact that "profit for its own sake" is not something that is a sole priority of mine.

You reach a point in your life where "legacy" takes precedence over other considerations. Assuming a man has made something of his life, what a man does once he has reached a certain threshold of self-awareness has less to do with the accrual of wealth and more to do with what kind of future he can provide—for his families, for his loved ones, for his circles of involvement, and for his immediate universe.

    I have perhaps learned that the hard way. Because for some reason none of us can explain, we all manage to fantasize that we're going to live forever. But God and Mother Nature have other ideas. And they even have gentle ways of reminding us that this time we spend here on Earth is both precious and short.

CHAPTER 40

HEARTFELT

There are three kinds of milestones that come to us all at some point or another. The first two are usually wonderful. The third is not but nonetheless wedges its way into this mosaic of living in ways that we must account for, as well as cope with when they come.

Birthdays, weddings, anniversaries, holidays, vacations, banquets, conventions, special events and celebrations of life—these are touchstones that promise delight and often stick out in memory. Then there are those landmark events such as promotions, honors, awards, degrees, mementos of achievement—all those special items for your bucket list that make you feel that your life has had meaning, that you've made a mark on this world.

The third kind of crucial event is not quite so nice, at least on the surface. And yet it is part of God's plan for us that difficulty is occasionally the teacher we so often take for granted. It arrives with the hard times we all experience in our lives. Financial crisis, personal loss, an accident or a catastrophic illness—the inevitable events of strife that make us stronger and test our mettle.

Health, as I have mentioned, has always been a paramount consideration in my family, especially given the matrix of autoimmune diseases we've had to deal with. I also have to note here that it is usually the male of the species that tends to dismiss these health challenges and trudge ahead. At least in the case of this male, that has been very much the case.

Still, there have been three major health events—serious illnesses— that prompted me to pause and reappraise my situation, my career and the conduct of my physical habits. And that has made all the difference.

I mentioned the discovery of my Type II diabetes during a "physical" right around the time of my 40th birthday in 1978. That event certainly motivated me not only to reappraise my personal habits, but also the breakneck pace of my daily life. It also jolted me into becoming more aware of what was actually going on with my body. So I sat up and took notice. Longevity is not necessarily an Iskandar family trait, and I for one hoped to reverse that rather dark tradition.

Part of the challenge of having a family history of autoimmune diseases has to do with what they call "concomitant conditions" or side-effect illnesses that often drop in with a broadside seemingly out of the blue. This is exactly what knocked me down on New Year's Eve in 1986.

Even though I had moderated my lifestyle during the previous eight years, 1986 had proven an exceptional year for me. I had recently been moved up to Geochemical Sciences Branch Chief at CRREL and had worked night and day on several projects, publishing 6 articles and 2 books in the same year with Magdi Selim and some of my other science associates. I was on a roll in my newfound "moonlighting" as a real estate owner, broker and entrepreneur, and I was working nights and weekends maintaining our rental properties. So, to say the least, I was pushing the envelope, logging in what had to be 90 hours a week for several months on end.

It wasn't surprising then, when it came time to usher in the New Year, that I asked Bonnie if she wouldn't mind if we went to bed early and brought 1987 in bright and early the next morning.

I remember having a half a glass of wine and a snack and actually falling asleep at least an hour before midnight only to wake up about 2:00 a.m. in the New Year with an extreme numbness on my right side that went from my neck all the way down to my fingertips. Bonnie's first impression was that my arm might have gone to sleep because I'd slept on it at a strange angle. I also thought she was probably right, so I shook it off for the moment and dropped back off to slumber for a while longer. Then in a few hours I awakened to realize that the immobility was still there, as were the numbness and loss of sensation.

I sat up and tried to make things work, and that's when the realization struck me: I had lost all mobility in my right arm! It was virtually frozen. I had experienced a stroke! (Happy New Year!)

A trip to a savvy chiropractor a couple of days later enlightened me to the fact that I had undergone a bout of *hemiplegia*, or hemispheric paralysis on one side that was brought on by what he believed was probably a *stress stroke*—the kind that is often related to extreme exhaustion often aggravated by a touch of hypertension (which I already had).

The running joke is that "Death is God's way of telling you to slow down." I was reminded of that little conundrum by this event. And some moderate therapy and lifestyle modifications seemed to take care of the rest. Eventually, I regained all feeling and mobility; so that was a relief. And for a year or two, it seemed at least that I might have learned my lesson.

Unfortunately, calamities come in strange ways—some without warning, others self-induced. And I unwittingly brought the next major event upon myself. That came in early October of 1988, and might well be described as an "occupational hazard." In this case, it was because I was getting involved in a kind of skilled labor that should have been left to experts—the fine art of plumbing.

One foible to which I plead guilty is that I prefer to do everything myself. I have been that way for a number of reasons. The primary driving force is monetary. I have always taken on new tasks in order to save money, something I call "elbow-grease economics." Another reason is simply my desire to learn new skills. I like a challenge, pure and simple.

Then again, there are certain things you should just let other people do. Plumbing, I came to learn the hard way, was one. Of all contract labor a property owner has to pay out, plumbing is probably the most expensive. Underrated, to say the least, good plumbers charge a great deal for the work they do, and they deserve every bit of it. If I didn't know that back in 1988, I know it now.

Back in the day, however, I was a jack-of-all-trades and thought that some simple plumbing tasks, at least, were something I could handle. Well…there are things called "learning experiences," and I was about to get a dose.

It came at a building I had bought and was managing called 10 Parkhurst Street (apartments). Until that event, I had always personally taken on the responsibility for building maintenance and repair—everything from trash removal to sheetrock. It was originally a way to save money. Not only was it a savings, it also enabled us to build equity by improving the property and encouraging long-term occupancy.

In the case of the "plumbing" repairs in question, it was a simple pipe repair job. Some rusted out pipe joints in between units had been clogging up drainage and needed to be replaced. So I was removing a few rotted out lead/iron pipes and replacing them with PVC parts that would be both cleaner and more cost effective. The job was simple enough. I had done it before.

Unfortunately, I had forgotten to ask the tenants the basic question that is first on any experienced plumber's list: whether or not they had used any Drano or caustic compounds to clear out the drains. (If that had been the case, an experienced plumber would have "rooted them out" and then continued with the switch.)

Instead, when I went into the very small crawlspace to make the replacement and uncoupled the "L" joint on the pipe, a flood of caustics, including lye, sodium hydroxide and other unsavory liquids came cascading down onto my abdomen and almost immediately started eating into my skin.

The sensory shock was instantaneous. The pain and irritation that followed were extensive, and I knew I was in big trouble. Despite the fact that it was burning through my skin, I refused to leave until I finished the job. At that point I then ran to my car and drove home as fast as I could, jumping into the shower to try to wash things off. It was my first instinct but it may have not been the best idea, because water and oils are supposed to be used on a chemical burn after the area has been isolated, and showers can tend to spread the caustics around.

It quickly caught my attention that the shower wasn't working and may have actually made matters worse. So I had Bonnie drive me to the local hospital in Lebanon, but they immediately informed us that their ER was not equipped to handle anything as complicated as a chemical burn like this. So we dashed over to Hanover and Dartmouth-Hitchcock Medical Center, one of the finest hospitals in New England.

In very short order they treated the attack and disfigurement on my skin, following that up as soon as humanly possible with a skin graft to repair the damaged tissue and stanch the spread of any further infection. It involved a transfer of tissue from my thigh to the damaged tissue and was fortunately successful, even though the entire recovery process took over a month during the hottest time of the year.

The official diagnosis of what was essentially an industrial accident was officially as follows: "Second and third degree burns from caustic soda and other chemicals." Technically it was much more pervasive than that.

The best way to describe what I went through during that time in October of 1988, would be to compare it to a MRSA infection, AKA the [flesh eating] *Methcillin-Resistant Staph Aureus* super-bacteria. No, it wasn't contagious, and fortunately the healing medications worked for me, but the effects were just as damaging and almost as slow to heal. The pain was excruciating and it was terrifying to watch the flesh just fall off my body. I am so thankful to have recovered but will always carry the scars.

Lessons learned: From that point in my life, I stopped trying to do every single thing myself. I certainly developed a healthy appreciation for those tradesmen and their skillset. And I finally admitted that my time might be better spent not trying to be the handyman for the world.

Frankly, I thought that was about as bad as things could get, and that perhaps I'd paid my dues in the health challenge department. Of course I hadn't. And 25 years later the real test would come—the worst challenge of my life with a surprising set of side-benefits.

There are major and minor medical events in everyone's life. And if we are both lucky and wise when we get older we pay attention to our bodies so we can avoid those "catastrophic illnesses" that become life-threatening later on.

Over the years I had gotten a few warning signs that I had to address such as a slight scare I had in 2009 with a large a deposit of fatty material on my left thigh (called a *lipoma)* that turned out to be benign. Of course, we had "the tumor" removed before it turned into something serious. And the good news about manifestations like that is the fact that they are so physically imposing that one would simply be ridiculous to ignore them.

Unfortunately, it is those systemic symptoms that we tend too often to disregard. They're not visible, and so we repress all thoughts of them. Especially when things are going well, we just don't want to sweat the small stuff. So that's what I was no doubt guilty of during the summer of 2014. Unfortunately, it caught up with me…in spades.

It was late September in the selfsame year, a beautiful early autumn. In so very many ways, I was at the top of my game. My finances were solid. I was truly enjoying the new "entrepreneurial" business lifestyle. My activities with ISTEB and other geoscience groups were still prolific enough for me to keep my hand in this gathering of the finest geoscience minds in the world. My family relations were solid (as solid as any large family can be). And I was even finding a little time for Bonnie and me to travel.

Granted, I had some warning signs that should have been a red flag to me, but as usual I tended to dismiss them as minor and continued to soldier on. I had noticed an increase in shortness of breath after very little activity. My blood cholesterol and triglycerides were high. In my *A1C Tests,* my blood sugar measurements could have been better. There seemed to be some indication of arterial blockage that might "eventually require surgery." The doctors had already admonished me to drop a few pounds and slow down my customary hectic pace. After all, by then I was 75. And in terms of recent generations, I was unofficially becoming the longest living member of my family.

Then in September 2014—before, during and after a long-planned trip to Japan— the bottom dropped out on my health. Even though I was starting to feel fatigued and knew my energies were flagging beforehand, I was determined to make the trip to Fuokoka, Japan to meet with the local facilities committee in preparation for the upcoming ISTEB convention to be held there in July of 2015. And I also wanted to share some insights into possible remedies to repair the ongoing nuclear calamity at Fukushima—that continued to plague the entire ecology of the Pacific Ocean since the Earthquake in 2012.

Once I had arrived in Fuokoka and met with our new ISTEB and ICOTBE officers and committees, the warning signs came early and often. Shortly after joining them, I found myself unable to keep up with the brisk pace they set on their walks, and often had to stop and catch my breath before I would be able to rejoin their activities.

Even though I got through the few days in Fuokoka with some difficulty, I really hit "the wall" going back through China when I was required to change planes in Shanghai's Pudong Airport. Even carrying light material, I had trouble wheeling along my own small bag. I began sweating heavily, and I had lost my strength. So, I asked a Porter to help me get a cart so that I could change planes successfully.

On the flight back to the US and our port of entry at New York, I began to realize that I was in serious trouble where my physical condition was concerned. The turning point came as we were coming in for a landing at JFK Airport. Just prior to deplaning I had started feeling completely enervated and kept having to stop to rest and catch my breath. Even waiting to get off the plane, I was having difficulty breathing and walking. Suddenly I was struck by the sickening awareness that I was no longer able to carry my luggage or even continue under my own power. Finally, I was forced to get into a wheelchair just to get through US customs.

By the time I had gotten back home in New Hampshire, Bonnie made sure that I went to our family doctor the next morning to address the issues of my virtual collapse and my apparently failing health. After undergoing a thorough examination, we were informed that I was experiencing a highly irregular heartbeat. And after taking the stress tests and ultrasound, I received the additional prognosis that I had a bad valve in my heart. So our medical team immediately scheduled me for open-heart surgery at Dartmouth-Hitchcock to replace it.

It should be surprising to no one who has ever undergone this kind of physical invasion, life saving thought it may be, that the post operative recovery is the worst part of it. There you are lying in your hospital bed, having basically just been cracked open from stem to stern and wondering if you're ever going to get back in control of your life. The pain of movement is pervasive, but the rudest awakening of all is the realization of just how fragile we are.

It was then that I was forced to face the finality of life. I still had so much to accomplish, and yet my time was running out. Predictably, the thoughts racing through my mind had me assessing our family, my heritage and the fact that I had a responsibility to others to put my house in order. And yet, for the first time in my life I felt powerless. In what amounted to a true reality check, Bonnie and I became convinced that we would do better living in a single-story residence and decided that we needed to buy a new house and move before I was released home for recovery.

I spent the next few weeks working feverishly on amending our family trust. I feared that this would be the final say in my legacy and that it had to be at least clarified from the previous drafts we had done. Simultaneously we were busy looking for a new house. Fortunately, we yet again experienced a kind of Divine intervention, and the second property we looked at was perfect for us. Without hesitation, we put in an offer and were met with approval very quickly. But truly, this was the easy part. The hardest part of this entire transition was my utter inability to effectuate any real personal control.

Unashamedly, I once again revert to my mantra that *Crisis Creates Opportunity.* It has never failed me in the past, and it favored me once again. And as far as my family dynamic was concerned, it proved to be the ultimate gauge of character. Crisis, I have found, brings out the best in people; and the very worst. And the best people always rise to the occasion.

This happened to be in spades during my surgery and recovery. The surgery went well, and there was not a real life-threatening moment (although anyone who has gone through open heart surgery will assure you that they all are). The toughest part was the recovery and the pain of limited mobility because your chest is cracked open very much like that of a chicken being boned. So that, when they pull you back together and sew you up, your every joint, rib, cartilage and sinew feel as if they have to retrofit and learn to work together all over again.

My surgery, my loss of stamina and certainly my loss of mobility for quite a few weeks left me for the first time since my childhood completely reliant upon others for the extension of all my needs. Certain people were there for me every single moment. My wife Bonnie was at my side 24-hours a day, 7-days-a-week, attending my every need…and more. She was the unifier. My old friend Magdi Selim was there for me (as I had been for him on three occasions), underscoring more than ever that "families and fraternity" are often formed outside the bonds of blood.

Perhaps the most redemptive experience of all for me and the very best moments I took away from all of this were the ways in which my daughters Amy and Niveen rose to the occasion, each with her own approach in seeing to it that every aspect of my estate, all my issues of legacy, property and family stability were addressed.

Although the two had gone along their own destined paths in life, Niveen and Amy had bonded rather pleasantly during the one-week gathering for my 75th Birthday Celebration, and had truly enjoyed one another's company. There was a kindredship of spirit there, and I for one had enjoyed bearing witness to it.

I had been able to spend a great deal of time with my daughter Amy during the majority of the time she was growing up. I regretted more than anything not being able to enjoy the same access to Niveen during her most formative years. The blessing of it was that the time we did spend together was a concentration of love and spirited good fun. I had always made myself available to her over the years and particularly relished in the fact that Niveen, as the older female sibling, had inherited that "achievement" gene so characteristic of certain members of my family.

Mature, stable, and an excellent student, Niveen had gone on to Ripon College in Wisconsin where she got a Master of Science degree in psychology (and is still on track to get her PhD). She didn't know all that much about my business dealings, but when she learned of my near-death experience and my open-heart surgery she was very quick to come to my side and contribute in any way I wished her to do.

With this very imminent "wake-up" call I received from the Universe, I was determined to see to it that my estate, my finances and my trusts were solidly locked down. And in that regard, I was both relieved and delighted to behold the almost innate cosmic rhythm my daughters seemed to find in working with one another.

Amy has a surprising sense of business acumen and possesses some uncanny street savvy instincts about how to cut to the chase quickly and get things done. I say that as the highest form of compliment. Amy is very no-nonsense, and people respect her for that. So she has proven herself to be very capable in the day-to-day operations of Lebanon Property Management, and nothing underscored that ability better than my period of convalescence. She immediately assumed the reins of service, liaison and operations, making sure that everything got done and that smart decisions were made without having to bother me with every minute detail.

One of the smartest decisions Amy made was to see, as I did, Niveen's innate skills of organization and managerial style balance the delicate social chemistry inside our own family. Niveen, with the instincts of a born leader, has displayed some remarkable people skills. Other people innately trust her. She projects a level of integrity laced with compassion so depictive of my own mother that, upon occasion, the resemblance is uncanny.

For that reason, both women agreed that Amy should be in charge of the business end of my estate while Niveen would act as the Trustee, all of us together determining how to protect the assets while making sure that the needs of my inheritors—children and grandchildren—would be appropriately addressed.

Anyone with a sizable estate to bequeath has to strike a delicate between revocable trusts, irrevocable trusts, living trusts and health trusts. The secret lies in appropriate variation—in minimizing tax liability and maximizing flexibility and freedom of access.

Between my wife and my two daughters, I have enjoyed watching the formation of a kind of business "dream team," one that has enabled me to recover with the realization that, as I move forward in health and a more balanced lifestyle, I have also set the future in place…and will be leaving it in some very capable hands.

Now, I'm fully mobile again. My heart rate is solid. I have kept my weight and my cholesterol in balance. And my A1C "sugar count" blood test is at an all-time low (7.2) range. I have more energy than I have had in quite some time. And my medical prognosis is rosy.

More than anything, my "trauma" has yet again proved to be a teacher. I not only learned more about myself, but also about my family and friends: who will be there when crises come…and who will not. The philosopher John Churchill Collins once wisely said: "In prosperity our friends know us. In adversity we know our friends." Those opposites also reveal things for what they are and people for who they are. And you can't put a monetary value on knowing such qualities as that.

So, for now I am recovered. My future prognosis is good. And I am more aware than ever of what really matters in life.

CHAPTER 41

LESSONS LEARNED: AN EXAMINED LIFE

The philosopher Aristotle once said: "An unexamined life is not worth living." I suppose this book has to have provided at least part of that examination.

Life itself is an object lesson. Virtually everything and everyone is your teacher. My memoir has perhaps disclosed some life lessons to me that I have hopefully been able to share with you. And yet as I review what I have taken away from the experiences dealt me, it is our interactions with other human beings where the teaching is the most indelible and the memories most vivid.

Reverting to my Coptic Christian roots, I have to note that all the stories and books, all the Psalms and Proverbs of the Old Testament were never fully understood until Jesus the Christ became divinity in man to give them both meaning and Infinite Light.

Our lives are filled with lessons: dozens in a day, hundreds in a week; thousands in a year. All we are required to do is recognize and accept them for the special gifts they are meant to be. What's more, it seems we have teachers every day—some of them deliberate, others unintended—each of them valuable in their way.

Our families teach us in so many ways, and yet strangers do as well. Those who are dropped into our daily discourse with some message we never expected, those who have challenged us, or tried to oppose us, or those who suddenly help us for reasons none of us could quite define—all have value in the most remarkable of ways.

From my mother Aida, I certainly learned to be clear and honest in my communication; but always with kindness and a true love for all things. She also taught me the fine art of frugality—how to save all my resources from energy to money. I also learned from her that hunger and deprivation are teachers, and to acknowledge them so that I may place greater value on all the days to come.

From my sister, Fayza, I was taught the blessings of forgiveness, generosity of spirit and unconditional love. From my father Karam I learned the gift of redemption, notably seeing him come together later in his life, once he had overcome his "addiction." From my siblings, especially the ones lost at such an early age, I learned the importance of living fully every day, loving and sharing time with those you care about with a sense of joy and urgency. From people like Nora Gunther, the Reverend Earl Parchia and the wonderful brewery manger in Milwaukee who pointed me in the direction of Madison, I learned that our lives are filled with (angelic) messengers who line the way to our destiny if we will but keep our senses open to the truth of it, and recognize it when it comes.

From Senator William Proxmire and other leaders who have influenced me even from the fringes of my life, I've learned that you are never too big or too small to make a positive impact on someone else…and that it is not only our opportunity but also our covenant as decent human beings to "pay it forward." (The secret comes with the fact that the more you do with your own life, the more you are able to make a positive impact on those around you).

From brilliant mentors like Dr. Keith Syers I learned, and I firmly believe, that there are "guides" brought into our experience to uncover for us the fulfillment of our dreams. If we will just stay on purpose and be open to those opportunities when they come, they will present themselves. From my dear friends like Magdi Selim and Hassan Moawad I have learned that families are not always bonds of blood but the intertwining of gold and steel of mutual achievement and higher minds that often make the best and longest lasting relationships. From colleagues such as Magdi Selim, Hassan Moawad, Hillary Inyang, Domy Adriano, Albert Page and other professional associates, I have learned that some friendships are fire tested and certain to last a lifetime.

From my wife Bonnie Iskandar, I have learned the meaning of true love, and that of a soul mate who always has my best interest at heart, with the integrity of caring for others that elevates everyone around her.

From my children—all of them, high achievers and troubled souls alike—I have learned to give what I have, and to love without condition or expectation. They've also taught me importance of keeping what I call "an open hand," to allow others to lead their own lives and find their own path without the need to judge. By the same token, I have also come to realize that one's first obligation is to oneself, and to draw the line where firm conduct and good morals are the ultimate standard against which all other decisions must be measured.

Finally, I have learned from the children of all my families that, at any stage, intermingling lives is a delicate art. It requires strength and patience, insight and kindness, and—above all else—the realization that nothing is perfect, and acceptance brings peace of mind.

From those who have opposed me or tried to block my efforts, in my life, I have learned the ultimate lesson from *Sun Tzu, The Art of War*. "You often learn more from your enemies than you ever do your friends, because they require you to reexamine your own motives and seek a better outcome."

In so many instances, I have learned that very few people mean to harm you; they are only acting from fear and a lack of information. No matter what your goals or visions from the outset, if you involve everyone in the process, you exponentially increase your chances to succeed. *Inclusion* is the secret here. Dreams are contagious. Embrace others with them early and often. Let them see what you envision. Let them share in the process, and you will have allies, compatriots and friends for the rest of your days.

Ultimately, I could probably fill this chapter with a thousand life lessons. But what I have learned in the strongest possible way is that there are really only a few to which we need to give our focus. To be sure, everyone's lists will vary. These, for better or worse are mine. And I would like to share them with you now.

- *Your time is the only Capital you bring into this life. Spend it wisely.* Successful people use their primary years for knowledge and skill building until about age twenty; their years of optimum productivity last until they are sixty or seventy (depending upon their health); and the final phase of a person's life is slower. Killing time kills your success. By choosing to make the most out of my time I was able to work two full time jobs and become successful. (I don't recommend two jobs for everyone at this stage of the game. But at least do this much. For at least two weeks, make a log of how you spend your hours and days and what you spend them on. You may not want to work two jobs, but how about learning a new language? Taking a new course? How about seeking a new degree? Or taking your hidden passion and developing an "avocation" from it? Very often hobbies are our hearts desires, and ultimately what gets us up in the morning.)

- *It is important to plant the seeds of success early in life and to continue nurturing them until you achieve fruition.* And be sure to do so with both integrity and intention. It has occasionally been noted that, "character is destiny." And what you focus on grows if for no other reason than this: Your energy is oxygen to your dreams.

- *Find your passion. Let it be the force that drives you and continually moves you toward each new horizon.* Embrace the unexpected, for it holds the key to all the good things coming your way. Be prepared to act when opportunity knocks and be ready to take a risk. Remember: if you believe in yourself, the risks are always slight, because determination is the deciding factor in everything you undertake.

- *Always be willing to start something new.* Doing so is the surest path to growth. Moving, changing careers, going into business for yourself, giving all to the pursuit of a dream—these are what make life worth living. They take you out of your comfort zone and drive you to excel. And leaving the "known" to pursue the unknown is the very best way to grow.

- *Never burn bridges or hold onto grudges.* Often those who oppose you now may become your allies later. So look for a common ground. And forgive; rise above every situation and make new friends as you do.

- *The strength of a person can be measured in their self-control.* When people lose their temper, they lose their leverage. So think before you act. You can't take back negative words and actions. And your best words and deeds define you. People will remember you best by your attitude and actions. Positive energy is power. So keep your hand on the switch.

- *Understand that Failure is a part of the process of Success.* Experiments fail so that we can learn from our mistakes and improve our strategies to accomplish our goals. Errors often frame the window to our focus.

- *I've always held true to the adage, "If at first you don't succeed try, try again."* It's a fact: Most failures occur on the threshold of success. So, the only real failure in life comes when you quit. Since that is the case, and I believe this wholeheartedly, make sure that you believe 100% in what you're doing. Then go after it with all your heart. The rest is just a matter of time…and application.

- *Always embrace the Law of Like Minds.* You attract exactly the people in your life that you deserve. So, surround yourself with those you want to emulate, those who can show you new ways to achieve what you really want. This is the *Law of Attraction,* pure and simple. It is important to have role models who will both encourage and challenge you to excel.

- *Spend your "Emotional Currency" wisely.* Next to your time, your *personal energy* is the most valuable commodity you have in life. It is precious and fragile. So spend it on the right things. (You'll know when you do because you'll come away energized by the experience.)

- *CRISIS CREATES OPPORTUNITY.* Very often, your best opportunities come when things appear to be at their worst. By remaining calm and centered you will soon learn to perceive the opportunity inside the crisis. These situations have yielded the highest rewards. So stay focused. Appraise the situation intelligently. And then act, quickly and decisively.

- *Diligently Do Due Diligence.* Always do your homework and take care to learn about any new ventures. You won't have perfect outcomes every time, but this is key to managing risk.
- *All Progress is made from the Inside/Out.* Build yourself. Build your team. Build your relationships. Set your game plan. And act. Yes, it's as simple as that.
- *Always look 10 years into the future.* Have a bold vision, and show the courage to follow it through. Establish short and long-term goals, and then create a specific plan to build them into a reality.
- *Be grateful and thank God for each day on this earth with appreciation for the abundance of life.* This, first and last, is the secret of experiencing joy and happiness; be content with what you already have. Be grateful for all the wondrous mysteries that await you. The Christian philosopher Meister Eckhardt once said: "If the only prayer you said was 'thank you,' that would be enough." What I have noticed throughout my life is this: When you are grateful, fear and sadness vanish, and abundance follows. And that is just the beginning.

This life is a gift—every moment. And now that I've reached that certain turning point when I have to consider what indeed will be next, I have come to realize that it is not an end but a beginning. In so many more ways than I might have imagined before my illness and recovery, I have realized that I have been granted a "life-extension" with an unlimited expiration date. It is a virtual palette of possibilities. It is a blank canvas, and I get to be the artist.

Since I have accomplished certain things in my life, and I've done a few things right, I find that I am empowered to pursue just about any new goals I like. The fun will come in the deciding, in creating a whole new agenda.

Let's just call it a New Bucket List, filled with some very good things…

CHAPTER 42

NEXT...A NEW BUCKET LIST

At this point, I'm glad that I've gotten to this point. However long the journey, no matter what we believe we will end up doing, practically nothing turns out exactly the way we thought it would when we began.

Many things have been better than I expected; others not as much. And yet now that I have come to this point in my life, getting my house in order is only one of a long list of things I would truly like to do.

We certainly have high hopes for Evenchance and the very modern, futuristic settlement that is being developed there—put together carefully and with both the inclusion and consent of everyone involved in this community of the future. Although its aims are modest and gradual, it contains designs and planning that go far into the future—50 years at least—and that is the kind of model I would like to see for others of its kind, a community built for generations to come by those with an eye to the future...and the heart to see it through.

I firmly believe that there are those of us who not only have the opportunity but also the moral obligation to create a better world both in our immediate neighborhoods and in the townships they combine to create.

I have been fortunate enough in my life by now to embrace the philosophy put forth in the Athenian Democracy of Pericles through a concept he called *Arête*. Liberally translated from the Greek, it means, "the cup that runneth over." The origin of this philosophy of abundance stems from the belief that a man who builds his wealth and plans his life in service of others can, after he has achieved his goals, become like a fountain whose bounty overflows into the lives of those around him. I have been blessed to be in a position to meet at least part of those criteria.

My goals and dreams to help generate a series of progressive environmentally coherent communities remain both a passion and a priority. And yet when it comes to what we refer to as "charitable causes," I turn homeward to Egypt, and what has become a nation in turmoil, and a Coptic Christian community under siege.

America is simply the most generous, big-hearted country in the world. People here give freely, prodigiously and very much from the heart to help those in need. It is also a nation with over 1.3 million officially recognized charities, many of which are very deserving and provide excellent services for everything from wounded veterans to children with muscular dystrophy. Still others are a bit top heavy—some of which list "administrative costs" that represent anywhere from 35% to 78% of the contributions fed into them. (We're not here to judge, but we are in a position to place our time and our money where we think they are spent most effectively.)

Over the years I have finally come to the conclusion that vetting my sources is not only important but also essential for survival—in any endeavor. That includes my contributions to charities. I think the old adage that says, "Charity begins at home," also points to one's need to have a total understanding of where his or her discretionary funds should be donated. In that tradition, I have become very involved with three favorite causes.

One is called The Littlest Lamb. It is an orphanage in Cairo specified for orphan Egyptian girls. It is run by a Coptic Christian family I have come to know personally.

It is a little known scandal in the Middle East that there are more than one million orphans wandering around the cities of Egypt, more than 60,000 of them in the streets of Cairo alone. You can just imagine what kind of fate awaits orphans in a Third World country, and the danger to young girls in a big city like Cairo is almost beyond comprehension.

The Littlest Lamb [www.littlestlamb.org] offers a comprehensive plan for housing, education, health care and social assimilation for about 250 young Egyptian girls. It is truly a well structured "home" to prepare these promising young women for life. Although, it is a Coptic Christian school and shelter, it offers safe-haven to Christian and Muslim girls alike, and has proven to be a model of restoration, preparation and development. So I'm quite proud of the association, and I have full confidence that my contributions are being well directed.

Another Egyptian charity to which I have devoted my time and energy is Care 4 Needy Copts [www.care4needycopts.org]. Care 4 Needy Copts (C4NC) was founded in 2007 when The Shepherd and Mother of Light, a humanitarian service group in Egypt invited key Egyptian Americans over on a special mission to witness the distressing environment within the Christian population in their native country. This visit ignited our passion to pursue the cause and advocate awareness in the US. The inhumane conditions and extreme poverty that some Copts are now forced to endure has prompted the formation of this organization, and finally it has gained some traction on both sides of the ocean.

It is a lightly held secret that Coptic Christians have been receiving the brunt of an Islamization of Egypt that has gone from gradual in the 1970s and 1980s to one of geometric progressions in the last 20 years, including a Coptic flight that has seen the Coptic Christian population diminish from one estimated at 22% of all Egyptians in 1955 to less than 10% in the year 2014…roughly a 55% drop in the last 30 years. The official evaluation of this "disappearance" is listed as unexplained. In truth, it is explained quite easily. Copts are, and have been, fleeing Egypt for decades. They are, more than ever, second-class citizens in the expanded radicalization of Islam, especially in Egypt in recent years.

This has become a particular concern since the year 2011 where we recently saw Coptic Churches, schools, charities, businesses and orphanages flagged with an "X" by the Muslim Brotherhood as targets for future sabotage and elimination.

This has been somewhat mitigated with the military-led revolution ousting President Mahamed Moursi in 2014 and the ascendancy of Egypt's new moderate president Abdel Fattah el-Sisi. But the street threats are still there, and Coptic Christians are still under pressure both in Egypt and other countries such as Syria, Yemen, Libya, Iraq and Sudan.

With that in mind, Care 4 Needy Copts has been set up to raise public awareness in Coptic and non-Coptic communities worldwide about the plight of Coptic Christians—especially to provide sustenance for as many families as possible through monthly financial support. These contributions often address their immediate needs such as food, shelter, clean water, sanitation, and more. C4NC also works to provide immediate medical assistance to the sick and injured as well as securing education and employment opportunities to help them get back on their feet.

Finally, I am in the process of setting up a "scholarship" fund for promising Egyptian students to enable them work on the research and initiatives to get them over to the USA to complete their higher education. Nowadays, at Egyptian Universities such as Cairo and Ain Shams, women—especially Coptic Christian women—are being denied the access to higher degrees once so readily available to them as recently as thirty years ago.

So, in my way, I'm providing a window of opportunity for worthy students so that they might be able to follow the academic path I once took and gain access to some of the same potentials that will open an entire realm of career horizons as well. Just as there were those who influenced me or helped me along the way, I hope to be able to "carry the torch" and bring others to this side of the ocean to find their futures here.

It's not as easy as it used to be. It is a different world today—more complicated and fraught with peril. There are more obstacles to progress for immigrants and transfers than the path that was once laid out for me several decades ago. Still, with the right effort and by understanding what might actually be accomplished, I remain hopeful in the belief that I am able to help others find what I have found: a journey to fulfillment and any soul's ultimate dream, to achieve great things and to prosper in the process.

Although I will quite likely not be around to see every wish fulfilled, I have complete confidence that my Trustee will pursue this with all her notable passion and see this goal fulfilled.

In that very small yet wonderful way, this is a journey that never ends. There's always another step to take. And around each corner, a dream…

"We are such stuff as dreams are made on," said the Bard. "And our little life is rounded with a sleep." I hope to find that sleep with the peace of mind that at least I did my best. That is all that anyone can do—and the last of our Agreements.

MILESTONES

A CHRONOLOGY OF EVENTS

No resume can summarize someone's life. And by the same token, even the most meticulous autobiography can leave a few things off that we would like to talk about. We also recognize that in creating a memoir, we are prone to hopscotch around in order to cover the important events of a moment. What we've done with this Chronology of events is actually to put them in a linear timeline, and give you references to points in the book where you might find them.

— Born in Cairo, Egypt - October 22, 1938.

— Graduated from *Dar El Tarbia/El Hadietha* Middle School—May, 1954.

— Graduated from *Shoubra* High School—June, 1957.

— Graduated with a Bachelor of Science Degree (BS) from Ain Shams University—1961.

— Worked for the Ministry of Agriculture (Soil Sciences Specialist)—1961-1968.

— Received Master of Science (MS) Degree in Soil Sciences from Ain Shams University—June 1967.

— Married to Marcelle Adib Ibrahim—April 17, 1966.

— Birth of Daughter Niveen Iskandar Karam. December 1, 1967.

— Published Masters Thesis: "The interrelationship between groundwater and soil salinity." Ain Shams University, Cairo, Egypt. 1967.

— Moved to America to (Milwaukee and) Madison, Wisconsin. October 10, 1968.

— Project Specialist at the University of Wisconsin Food Research Institute – November of 1968.

— Enrolled at the University of Wisconsin (Madison) as a Research Assistant and PhD candidate in the Department of Soil Sciences—September 1969.

— Co-Author. [J.M. Goepfert, I.K. Iskandar, and C.H. Amundson.] "Relation of the heat resistance of salmonellae to the water activity of the environment." *Applied Microbiology.* 1970.

— Birth of son George Iskandar Karam Iskandar. Madison, Wisconsin. January 19, 1971.

— Published "Pedogenetic significance of lichens," [with Keith Syers and I.K. Iskandar] as a Chapter in *The Lichens,* 1971. Vernon Ahmadjan and Mason Hale, editors.

— Initial publication of PhD Thesis: "The role of lichens in rock weathering and soil formation and mercury in sediments." I.K. Iskandar. University of Wisconsin. 1972.

— Co-author [I.K. Iskandar, J.K. Syers, L.W. Jacobs, D.R. Keeney, and J.T. Gilmour]. "Determination of total mercury in sediments and soils," as an article in *The Analyst.* 1972.

— Graduated – PhD in Soil Sciences from the University of Wisconsin (Madison) in June 1972.

— Lecturer; College of Environmental Sciences, University of Wisconsin, Green Bay. 1973.

— Co-Author of the Abstract: "Distribution and background levels of mercury in sediment cores from selected Wisconsin lakes. Water, Air and Soil Pollution." [J. Keith Syers, I.K. Iskandar, and D.R. Keeney.] 1973.

— Postdoctoral Project Associate; Department of Soil Science, University of Wisconsin. 1973-1975.

— Divorced in Green Bay, Wisconsin. November, 1973.

— Co-Author of the Article: "Concentration of heavy metals in sediment cores from selected Wisconsin lakes." [I.K. Iskandar and D.R. Keeney.] *Environmental Sciences and Technology.* 1974.

— Co-Author of the Article: "Sediment characteristics in the vicinity of the Pulliam Power Plant, Green Bay, Wisconsin." [J.M. Pezzetta and I.K. Iskandar, Co-Authors] *Environmental Geology.* 1975.

— Assumed position as Research Chemist; Earth Sciences Branch, U.S. Army Cold Regions Research and Engineering Laboratory [CRREL], Hanover, New Hampshire. February. 1975.

— Author and Correspondent: "Urban waste as a source of heavy metals in land treatment," In Proceedings International Conference on Heavy Metals in the Environment, Toronto, Ontario, Canada. Oct. 1975.

— Army Science Conference Award for outstanding publication. 1976.

— Purchased my first home on 3 Kinne Street in Lebanon, New Hampshire. August, 1976.

— Married to my best friend and soul mate Bonnie Palmerston in Lebanon, New Hampshire. October 15, 1976.

— Purchased my second property— a five unit (Barn Conversion) Apartment Unit on 94 Mascoma Street. September. 1976.

— Contributor/Co-author [R.P. Murrmann and I.K. Iskandar]. "Land treatment of waste-water-case studies of existing disposal systems at Quincy, Washington and Manteca, California." In *Land as a Waste Management Alternative* Ann Arbor Science. 1977.

— Author [I.K. Iskandar]. "The effect of wastewater reuse in cold regions on land treatment systems." *Journal of Environmental Quality.* 1978.

— Co-Author and Contributor [H.M. Selim and I.K. Iskandar] "Nitrogen behavior in land treatment of wastewater—A simplified model." In Proceedings, Vol. 1. *International Symposium on Land Treatment of Wastewater.* (First major collaboration with Magdi Selim. One of ten position

— papers and abstracts in 1978 alone.) Aug, 20-25, 1978. Hanover, New Hampshire.

— Diagnosed with Diabetes Type 2. (Insulin dependent) Hanover, New Hampshire. October, 1978.

— Attained my New Hampshire Real Estate Brokers License in August of 1980.

— Co-author [I.K. Iskandar and J.K. Syers]. Effectiveness of land treatment for phosphorus removal at Manteca, CA. *Journal of Environmental Quality.* 1980.

— Bonnie gives birth to our daughter Amy Iskandar on August 27, 1981.

— Editor [I.K. Iskandar]. *Modeling Wastewater Renovation — Land Treatment.* (First Published Book. A compilation of studies that still serve as a bible of wastewater management and renovation.) John Wiley & Sons, New York, NY. 1981.

— Co-author and Editor [Iskandar, I.K. and H.M. Selim]."Validation of a Model for Predicting Nitrogen Behavior in Slow Infiltration Systems," *Modeling Wastewater Renovation – Land Treatment.* John Wiley & Sons, New York, New York. 1981.

— Author and Initiator of an official CRREL proposal to the EPA Superfund to conduct a "Frozen Ground Pilot Study," showing permafrost technologies as a possible *pollution solution* for the environment. Funding Awarded. 1984.

— Co-Author/Collaborator [M.C. Amacher, J. Kotuby-Amacher, M.F. Hashim, I.K. Iskandar, and H.M. Selim]. Reactions and Transport of Cr (VI) in soils. Presented at the Workshop on "Mechanisms of Ion Transport in Soils." Zurich, Switzerland. May 20-23, 1985.

— Commendation Award from the Department of the Army for Outstanding Performance Rating as a Supervisory Research Physical Scientist. 1986.

— Chief of Geochemical Sciences. Branch Chief: US Army Cold Regions Research and Engineering Laboratory (CRREL). Hanover, New Hampshire. 1986.

— Co-Author and Collaborator [I.K. Iskandar, L. Perry, T.F. Jenkins, and J.M. Houthoofd], "Artificial freezing for treatment of contaminated soils

— A pilot study." In Proceedings of the 12th Annual Research Symposium, Cincinnati, Ohio, April 21, 1986.

— First Major Health Crisis - Suffered a stroke – *hemiplegia* (hemispheric paralysis) at 2 a.m. on New Year's morning, 1987. A wake-up call about my health requiring some intelligent lifestyle modifications, and learning to pace myself.

— Author and Editor of Article. [I.K. Iskandar]. "Ground freezing controls hazardous waste." *The Military Engineer.* Alexandria, Virginia. 1987.

— Certificate of Achievement for outstanding support in addressing under representation at CRREL of Women and Minority Scientists and Engineers. 1987.

— Co-Author/Collaborator [M.C. Amacher, H.M. Selim and I.K. Iskandar]. "Kinetics of Chromium (VI) and Cadmium retention in soils; a nonlinear multi-reaction model. *Soil Science Society of America Journal.* Madison, Wisconsin. 1988.

— Department of the Army Research and Development Award for Achievements in Geochemical Sciences—for outstanding performance in establishing site assessments and environmental reclamation in Alaska. Specifically for establishing a scientific basis for hazardous waste sludge treatment by applying frozen ground technologies. 1988.

— Second Major Health Crisis – Work accident involving an attack of caustic solutions on my stomach and abdomen. "Second and third degree burns from caustic soda and other chemicals." A severe burning and penetration of the skin similar to a MRSA attack. (Required recovery time, one month.) October, 1988.

— Co-Author [O.A. Ayorinde, L.B. Perry and I.K. Iskandar]. 1989. "Use of Innovative Freezing Technique for In-Situ Treatment of Contaminated Soils in Proceedings." 3rd International Conference on New Frontiers for Hazardous Waste Management. Pittsburgh, Pennsylvania. September 10-13, 1989.

— Purchased Stone Farm Apartment Complex. (36 two-bedroom units built in a Condominium format.) Lebanon, New Hampshire. 1989.

— Co-Author/Collaborator: [E. Chamberlain, I.K. Iskandar, and S.E. Hunsicker]. "Effect of Freeze-Thaw Cycles on the Permeability and Macrostructure of Soils." Proceedings International Symposium on Frozen Soil Impacts on Agricultural, Range, and Forest Lands. Spokane, Washington. March 21-22, 1990.

— Organized Workshop on Engineering Aspects of Metal Waste Management as part of the International Conference on Metals in Soils, Water, Plants and Animals. Orlando, Florida. April-May, 1990.

— Helped CRREL officially expand its Research and Development in Fairbanks Alaska, primarily to provide ($12 million government funded) application of frozen ground and environmental restoration technologies for the FAA and the US Air Force. 1990.

— The beginning of the concept of ISTEB. As an ad hoc committee member addressing the Soil Science Society, first put forth the future necessity for a Scientific and Professional Society. September, 1991.

— Editor [I.K. Iskandar and H.M. Selim]. *Engineering Aspects of Metal- Waste Management*. Lewis Publishers, Inc., Chelsea, Michigan. 1992.

— My purchase of Shipyard, Galleria in Hilton Head, South Carolina and an office complex in Fort Lauderdale, Florida. (Significant because it was our introduction to the tax advantages inherent in the 1031 Exchange.) 1992.

— Organizer/Coordinator. Workshop on Remediation of Soils Contaminated with Metals as part of the 2nd International Conference on the Biogeochemistry of Trace Elements, Taipei, Taiwan. 1993.

— CRREL Exceptional Performance Award. Hanover, New Hampshire. 1994.

— Co-Founder [I.K. Iskandar, D.C. Adriano, W. Wenzell, and A.L. Page]. International Society of Trace Elements Biogeochemistry (ISTEB). Lebanon, New Hampshire. 1994.

— Co-author [A.J. Palazzo and I. K. Iskandar]. "Use of Sewage Sludge on Park and Recreational Lands in Sewage Sludge: Land Utilization and the Environment." *SSSA Magazine,* Madison, Wisconsin. 1994.

— CRREL Exceptional Performance Award. Hanover, New Hampshire. 1995.

— Co-Organized International Workshop on Cold Regions Contaminant Hydrology, Anchorage, Alaska. (A seminal workshop and convention on permafrost and frozen ground technology.) August 22-23, 1995.

— Editor [D.C. Adriano, I.K. Iskandar and L. Murarka]. *Groundwater contamination.* Northwood, England, Science and Technology Letters. 1995.

— Co-Author [K. Mobley, I. K. Iskandar and H. M. Selim]. "Use of chelating agents for soil-metal decontamination." *Proceedings, Contaminated Soils – 3rd International Conference on the Biogeochemistry of Trace Metals.* Paris, France. May, 15-19, 1995.

— Exceptional Performance Rating and Performance Award, CRREL 1996.

— Editor/Coordinator [I.K. Iskandar, S.E. Hardy, A.C. Chang and G.M.Pierzynski]. Proceedings, Fourth INTERNATIONAL Conference on Biogeochemistry of Trace Elements, Clark Kerr Campus, Berkeley, California. 1997.

— Editor/Co-organizer. [I.K Iskandar.,E.A.Wright, F.K.Radke, B.S.Sharrat, P.H.. Groenevelt and L.D.Hinzman]. CRREL Report - Proceedings, International Symposium on Physics, Chemistry and Ecology of Seasonally Frozen Soils, Fairbanks, Alaska, June 10-12, 1997.

— Editor [I.K. Iskandar and D.C. Adriano]. *Remediation of Soils Contaminated with Metals.* Northwood, England, Science and Technology Letters. 1997.

— Co-Author [H.M. Selim, I.K. Iskandar and M.C. Amacher]. 1997. "Modeling the reactivity and transport of copper in soils." Proceedings Fourth International Conference on the Biochemistry of Trace Elements, June, 1999, Berkeley, CA.

— Co-Author [H.I. Inyang, I.K. Iskandar, and F.M. Parikh]. "Waste Containment Barriers; Physico-Chemical Interactions." *Encyclopedia of Environmental Analysis and Remediation, Volume 2.* John Wiley & Sons, New York, NY, 1997.

— Co-Author [H.I. Inyang, H.Y. Fang, I.K. Iskandar and M.R. Choquette]. 1997. Chemical and Mineralogical Analysis of Clayey Barrier Materials. *Encyclopedia of Environmental Analysis and Remediation, Volume 8.* John Wiley & Sons, New York, NY, 1997.

— Co-Chairman. Editor of the Anthology. [I.K. Iskandar, S.E. Hardy, A.C. Chang, G.M. Pierzynski]. *Fourth International Conference on the Biogeochemistry of Trace Elements.* Berkeley, California. 1997.

— Editor [D. C. Adriano, Z. Chen, S. Yang and I. K. Iskandar]. *Biogeochemistry of Trace Elements, Science Reviews.* Northwood, England. 1997.

— Co-Author/Editor [I.K. Iskandar and F.H. Sayles]. "Ground freezing for containment of hazardous waste: Engineering Aspects," *International Symposium on Physics, Chemistry and Ecology of Seasonally Frozen Soils,* Fairbanks, Alaska. June 10-12, 1997.

— Founded Lebanon Property Management. 18 Bank Street. Lebanon, New Hampshire. 1995.

— Distinguished Visiting Research Professor at The University of Massachusetts, Lowell, Massachusetts. Sept. 1997 [October 2000].

— Co-Chairman. Fourth International Symposium on Environmental Geotechnology and Global Sustainable Development. Boston, Massachusetts. 1998.

— Fellow, Soil Science Society of America. 1998.

— Fellow, American Society of Agronomy. 1999.

— Contributing Member, Organizing Committee, Fifth International Conference on the Biogeochemistry of Trace Elements. Vienna, Austria. 1999.

— Co-Organized and Moderated the Feature Symposium: "Bioavailability, Flux and Transfer of Trace Elements in Soils and Soil Components." Held during the 5th International Conference on the Biogeochemistry of Trace Elements, Vienna, Austria. 1999.

— Editor. [H. Magdi Selim and Iskandar K. Iskandar]. *Fate and Transport of Heavy Metals in the Vadose Zone.* Lewis Publishers. New York, NY. 1999.

— Editor [S.A. Grant and I.K. Iskandar]. *Contaminant Hydrology/Cold Regions Modeling.* Lewis Publishing. CRC Press. Boca Raton, Florida. 1999.

— Voted to be a Vice-President. International Society of Trace Elements Biogeochemistry. 1999 [2001].

— Sale of Shipyard Galleria and other properties combined with the purchase of 330-acre plot of land called Sleeper Village (and the office complex at 30 Airport Road)— through the exercise of the 1031 Exchange—Lebanon, New Hampshire. 1999.

— Editor [I.K. Iskandar]. *Environmental Restoration of Metals Contaminated Soils,* Ann Arbor Press. 2000.

— Editor [P.M. Huang, and I. K. Iskandar]. *Soil and Groundwater Pollution and Remediation: Asia, Africa and Oceania.* Lewis Publishers. CRC Press. Boca Raton, Florida. 2000.

— Honorary Member, International Society of Trace Elements Biogeochemistry (ISTEB). 2001 [2003].

— Chairman and President of the ISTEB International Conference in Uppsala, Sweden. June 15-19, 2001.

— Editor [I.K. Iskandar]. *Environmental Restoration of Metals-Contaminated Soils.* Lewis Publishers. CRC Press. Boca Raton, Florida. 2001.

— Instituted a proposal for a "Futuristic Planned Community" for Sleeper Village in 2002. Initiated Proposition 20 as an approval referendum on the community (also in 2002), which was defeated by 121 Votes. (We shortly sold the entire "undeveloped" 330-acre tract of land for a 500% profit.)

— Purchased 45 Lyme Road commercial building in Hanover, New Hampshire (on the Reverse 1031 Exchange). Negotiated a 3-way transaction that required some very sophisticated businesses and financial, and a leap of faith to boot. Result: a 55% profit on a large income-generating business complex. 2005.

— Establishment and Development of Evenchance Village, A Visionary Community of 180 Acres along the coast of Lake Mascoma dissecting West Lebanon, Lebanon and Enfield, New Hampshire. 2008.

— The "Alex Iskandar" 75th Birthday Celebration and first ever Iskandar Family Reunion. Orlando, Florida. October 15 thru 25, 2013.

— Author/Editor. A Personal Memoir. *Perfect Prosperity.* Fideli, Publishing. Bloomington, Indiana. July. 2015.

— Co-Host and Treasurer – ISTEB 16th World Conference of the International Society of Trace Elements Biogeochemistry. Fuokoka, Japan. July 12-15, 2015.

ACKNOWLEDGEMENTS

I am deeply grateful to my wife, Bonnie Iskandar for encouraging me to complete this book and for her unconditional love and support these past forty years. I also thank my mother Aida Barsoum for her enduring wisdom, kindness, and love. I wish to thank Dr. Boutros Yousef for saving my life when I was a child and who provided me with vital information about my father's lineage, and the late Bishop Earl Parchia SR., D.D. for hosting me at his home for the first two weeks after I arrived in the United States. I am also grateful to my cousin, Nabil Aziz Barsoum and my nephew, Maged Haddad who provided me with detailed information about my mother's lineage. To my brother Saied Iskandar, who reviewed this manuscript and provided valuable insights into our family of which I was previously unaware, I offer a heartfelt "thank you." Additionally, I would like to express my deep appreciation to my daughters Niveen and Amy for their commitment to this project, for their encouragement to pursue it, and for providing critical review and constructive comments at just the right times.

Thanks also to my professor, the late Keith Syers, who passed away in Thailand in July 2011. I am fortunate to have had such fantastic guidance from friends and mentors such as Dr. Syers and so many other professors and teachers who have helped shape both my scholarship and my approach to the intelligent application of this mission called science. I would also like to acknowledge my many colleagues such as Drs. Magdi Selim, Hassan Moawad, Domy Adriano, Albert Page, Steve Grant, and Hilary Inyang—brilliant minds, dedicated professionals and good friends with whom I have shared so many projects, research papers and publications. My sincere appreciation also goes to Drs. Harlan (Ike) McKim and the late Dwayne Anderson who selected me among many applicants to work at the US Army Cold Regions Research Engineering Laboratory (CRREL). And many thanks go to my good friend, Mr. Stavros Kritikos, and the many others who have influenced my life.

Also, I would like to recognize the International Society of Trace Elements Biogeochemistry (ISTEB) for its extraordinary work in the field of environmental science. It has been both an honor and a pleasure to have originally presided over this organization, and to see it expand and progress to the level its original founders envisioned.

Finally, I would like to thank Mr. Robert Joseph Ahola for his help in writing and designing this book. I am thankful to all those that played a part in this journey.

ABOUT THE AUTHOR

Iskandar K. Iskandar, PhD

Dr. Alex Iskandar's saga, *Perfect Prosperity*, crosses continents, oceans and generations to become what has to qualify as a classic success story that could easily be lifted out of the pages of Horatio Alger.

Born in the slums of Cairo, Egypt in the late 1930s, young Iskandar K. Iskandar worked his way out of an old world caste system that punishes initiative and eventually came to America in 1968, imbued with an iron sense of purpose that could not be denied.

A PhD in the rare field of Soil Science, Alex is a true Renaissance man who has been able to apply his prodigious gifts for pioneering scientific insights and his talents for savvy futuristic real estate development, and meld them into exceptional outcomes that have made a major difference in the world around us.

An initiator of groundbreaking global soil science discoveries, Dr. Iskandar has co-authored and anthologized 13 different published books on everything from *Contaminant Hydrology [Cold Region Modeling]* to *Modeling Wastewater Renovation,* as well as a series of (80) different abstracts and award winning papers that have helped change the world we live in. He is also the kind of broker and investor who is capable not only of brilliant real estate leveraging but also of generating a vision for ideal communities that will last 100 years into the future.

As his first mainstream work, *Perfect Prosperity* is Alex Iskandar's personal memoir and more. It is also his way of shining the light on a winning philosophy as a means of providing a role model in which others might find inspiration, motivation and delight.

www.ingramcontent.com/pod-product-compliance
Lightning Source LLC
Chambersburg PA
CBHW052012070526
44584CB00016B/1720